Achieve

AF767763

KS2 SATS MATHS

REVISION WORKBOOK

MATHEMATICS

EXPECTED STANDARD

YEAR 6

AGE 10–11

TREVOR DIXON

SARAH-ANNE FERNANDES

hachette LEARNING

Hachette UK's policy is to use papers that are natural, renewable and recyclable products and made from wood grown in well-managed forests and other controlled sources. The logging and manufacturing processes are expected to conform to the environmental regulations of the country of origin.

To order, please visit www.HachetteLearning.com or contact Customer Service at education@hachette.co.uk / +44 (0)1235 827827.

ISBN: 978 1 03602 280 8

First published in 2015

Second edition published in 2018

This edition published in 2026 by Hachette Learning (a trading division of Hodder & Stoughton Limited),

An Hachette UK Company

Carmelite House

50 Victoria Embankment

London EC4Y 0DZ

www.HachetteLearning.com

The authorised representative in the EEA is Hachette Ireland, 8 Castlecourt Centre, Dublin 15, D15 XTP3, Ireland (email: info@hbgi.ie)

Impression number 10 9 8 7 6 5 4 3 2 1

Year 2030 2029 2028 2027 2026

Authors: Trevor Dixon and Sarah-Anne Fernandes

Educational Adviser: Steph King

Series Editor: Sarah-Anne Fernandes

Accessibility Reviewer: Vivien Kilburn

Illustrations by Ann Paganuzzi

Typeset in India

Printed and bound in Great Britain by Bell & Bain Ltd, Glasgow

A catalogue record for this title is available from the British Library.

FSC™ www.fsc.org MIX Paper | Supporting responsible forestry FSC™ C104740

Contents

Welcome to Achieve Mathematics: The Expected Standard – Revision Workbook

In this book you will find lots of practice and information to help you achieve the expected standard in the Key Stage 2 Mathematics tests.

About the Key Stage 2 Mathematics Tests

The tests will take place in the summer term in Year 6. They will be done in your school and will be marked by examiners – not by your teacher.

There are three papers to the tests:

Paper 1: Arithmetic – 30 minutes (40 marks)

- These questions assess confidence with a range of mathematical operations.
- Most questions are worth 1 mark. However, 2 marks will be available for long multiplication and long division questions.
- It is important to show your working – this may gain you a mark in questions worth 2 marks, even if you get the answer wrong.

Papers 2 and 3: Reasoning – 40 minutes (35 marks) per paper

- These questions test mathematical fluency, solving mathematical problems and mathematical reasoning.
- Most questions are worth 1 or 2 marks. However, there may be one question worth 3 marks.
- There will be a mixture of question types, including multiple-choice, true/false or yes/no questions, matching questions, short responses such as completing a chart or table or drawing a shape, or longer responses where you need to explain your answer.
- In questions that have a method box, it is important to show your method – this may gain you a mark, even if you get the answer wrong.

You will be allowed to use: a pencil/black pen, an eraser, a ruler, an angle measurer/protractor and a mirror. **You are not allowed** to use a calculator in any of the test papers.

Test techniques

Before the tests

- Try to revise little and often, rather than in long sessions.
- Choose a time of day when you are not tired or hungry.
- Choose somewhere quiet so you can focus.
- Revise with a friend. You can encourage and learn from each other.
- Read the 'Top tips' throughout this book to remind you of important points in answering test questions.
- Make sure that you know what the bold key words mean.
- For longer questions, remember to practise showing your working: this could help you gain extra marks in the tests.

During the tests

- READ THE QUESTION AND READ IT AGAIN.
- If you find a question difficult to answer, move on; you can always come back to it later.
- Always answer a multiple-choice question. If you really can't work out an answer, try to think of the most sensible response and read the question again.
- Check to see how many marks a question is worth. Have you written enough to 'earn' those marks in your answer?
- Read the question again after you have answered it. Make sure you have given the correct number of answers within a question, e.g. if there are two boxes for two missing numbers.
- If you have any time left at the end, go back to the questions you have missed.

Where to get help

- **Pages 8–12** practise number and place value.
- **Pages 13–22** practise number – addition, subtraction, multiplication and division.
- **Pages 23–31** practise number – fractions, decimals and percentages.
- **Page 32** practises ratio and proportion.
- **Pages 33–35** practise algebra.
- **Pages 36–44** practise measurement.
- **Pages 45–51** practise geometry – properties of shapes.
- **Pages 52–57** practise geometry – position and direction.
- **Pages 58–63** practise statistics.
- **Pages 64–68** practise problem solving and reasoning.
- **Pages 69–70** provide definitions of all the key words.
- **Pages 71–73** provide the answers to the 'Try this' questions.
- **Inside back cover** contains a revision checklist to help you keep track of your progress.

How to use this book

1 ***Introduction*** – each content strand in the mathematics National Curriculum has been broken down into smaller topics. This introduction tells you what you need to be able to do for this topic.

2 ***What you need to know*** – summarises the key information for the topic. Words in bold are key words and those in lilac are also defined in the glossary at the back of the book.

3 ***Let's practise*** – a practice question is broken down in a step-by-step way to help you to understand how to approach answering a question and get the best marks that you can.

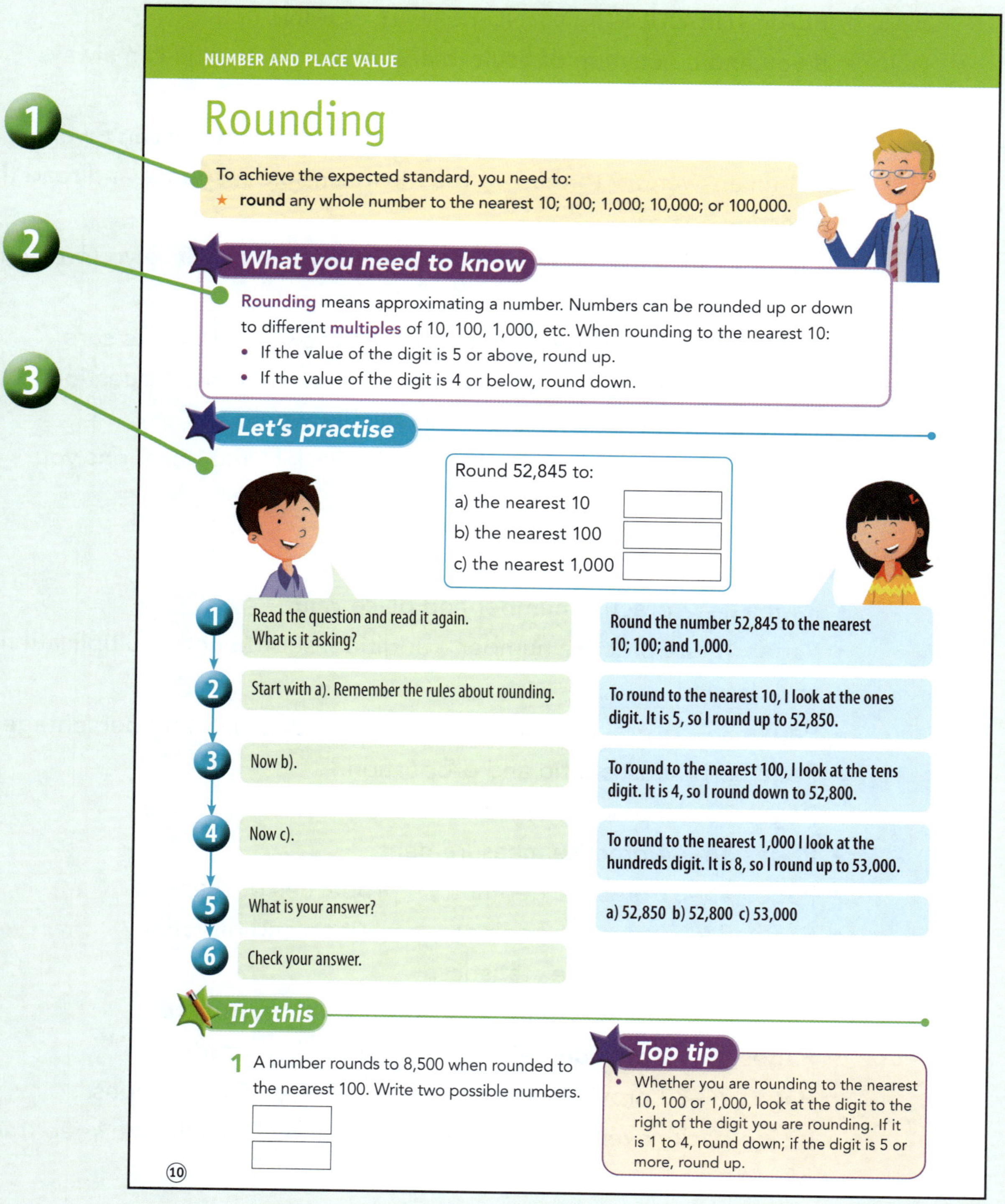

NUMBER AND PLACE VALUE

Rounding

To achieve the expected standard, you need to:

★ **round** any whole number to the nearest 10; 100; 1,000; 10,000; or 100,000.

What you need to know

Rounding means approximating a number. Numbers can be rounded up or down to different **multiples** of 10, 100, 1,000, etc. When rounding to the nearest 10:

- If the value of the digit is 5 or above, round up.
- If the value of the digit is 4 or below, round down.

Let's practise

Round 52,845 to:

a) the nearest 10

b) the nearest 100

c) the nearest 1,000

1 Read the question and read it again. What is it asking? — Round the number 52,845 to the nearest 10; 100; and 1,000.

2 Start with a). Remember the rules about rounding. — To round to the nearest 10, I look at the ones digit. It is 5, so I round up to 52,850.

3 Now b). — To round to the nearest 100, I look at the tens digit. It is 4, so I round down to 52,800.

4 Now c). — To round to the nearest 1,000 I look at the hundreds digit. It is 8, so I round up to 53,000.

5 What is your answer? — a) 52,850 b) 52,800 c) 53,000

6 Check your answer.

Try this

1 A number rounds to 8,500 when rounded to the nearest 100. Write two possible numbers.

Top tip

- Whether you are rounding to the nearest 10, 100 or 1,000, look at the digit to the right of the digit you are rounding. If it is 1 to 4, round down; if the digit is 5 or more, round up.

10

4 ***Try this*** – this is where you get the chance to practise answering questions for yourself. There are a different number of questions for each topic.

5 ***Top tips*** – these give you further reminders about answering test questions or help you to understand a tricky topic.

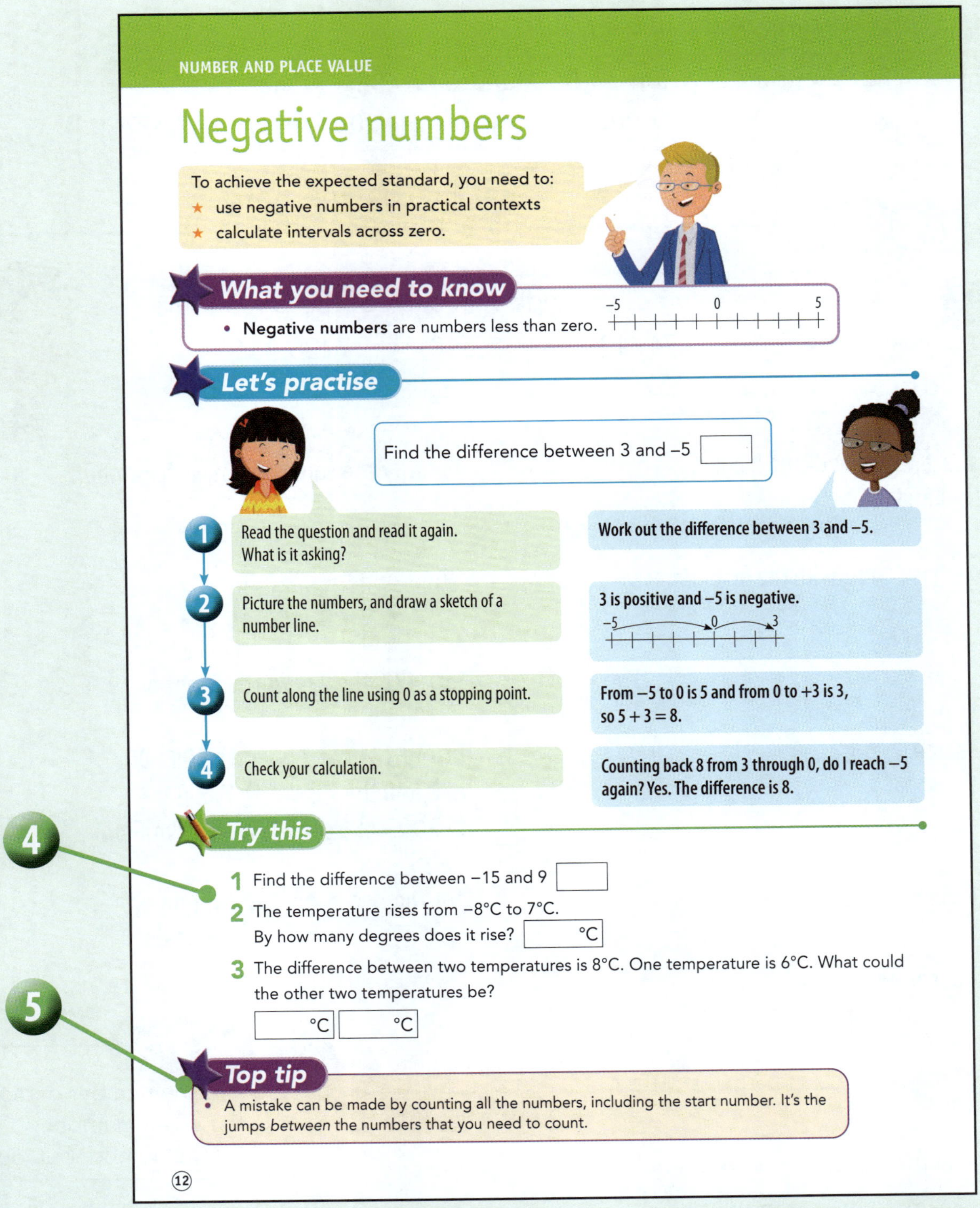

NUMBER AND PLACE VALUE

Negative numbers

To achieve the expected standard, you need to:

- ★ use negative numbers in practical contexts
- ★ calculate intervals across zero.

What you need to know

- **Negative numbers** are numbers less than zero.

Let's practise

Find the difference between 3 and –5 ☐

1 Read the question and read it again. What is it asking?	Work out the difference between 3 and –5.
2 Picture the numbers, and draw a sketch of a number line.	3 is positive and –5 is negative.
3 Count along the line using 0 as a stopping point.	From –5 to 0 is 5 and from 0 to +3 is 3, so 5 + 3 = 8.
4 Check your calculation.	Counting back 8 from 3 through 0, do I reach –5 again? Yes. The difference is 8.

Try this

1 Find the difference between –15 and 9 ☐

2 The temperature rises from –8°C to 7°C.
By how many degrees does it rise? ☐ °C

3 The difference between two temperatures is 8°C. One temperature is 6°C. What could the other two temperatures be?
☐ °C ☐ °C

Top tip

- A mistake can be made by counting all the numbers, including the start number. It's the jumps *between* the numbers that you need to count.

12

Place value of whole numbers

To achieve the expected standard, you need to:

★ know the **place value** of digits in whole numbers up to 1,000,000 and begin to know the place value of digits in whole numbers up to 10,000,000.

What you need to know

- The **place value** of each digit in a number depends on its position.
- One million has six zeros: 1,000,000.
- Ten million has seven zeros: 10,000,000.

Example:	Ten thousands		Thousands		Hundreds		Tens		Ones	
	40,000	+	7,000	+	300	+	20	+	7	= 47,327

Let's practise

What is the value of the digit **7** in each of these numbers?

1,472   749,121   7,540   7,123,456

________ ________ ________ ________

1 Read the question and read it again. What is it asking?

Identify the value of 7 in each number.

2 Be systematic. Start with the first number.

Millions	Thousands					
M	Hth	Tth	Th	H	T	O
			1	4	7	2

In 1,472, the 7 is worth 7 tens or 70

3 Do the same for the other numbers.

In 749,121 the 7 is worth 700,000
In 7,540 the 7 is worth 7,000
In 7,123,456 the 7 is worth 7 million

4 What is your answer?

70; 700,000; 7,000; and 7 million

5 Check your answer.

Top tips

- Count the number of digits in the whole number.
- Read the number out loud.

Try this

1 What is the value of the **4** in each of these numbers?

1,472   749,121   7,540   163,452   4,265,279

________ ________ ________ ________ ________

Comparing and ordering whole numbers

To achieve the expected standard, you need to:

- ★ use the place value of digits in whole numbers to compare and order numbers up to 1,000,000
- ★ begin to work with numbers up to 10,000,000.

What you need to know

- You can use the less than symbol (<) or greater than symbol (>) to compare numbers.
- Understand place value to compare and order numbers up to 1 million.

Let's practise

Write these numbers in order, starting with the **largest**.

844,202 560,342 2,152,889

[ ] [ ] [ ]

largest **smallest**

1 Read the question and read it again. What is it asking?

Write the numbers in order, with the largest number first.

2 Think of the value of each number. Do this by looking at the columns the digits are in.

Millions	Thousands					
M	Hth	Tth	Th	H	T	O
	8	4	4	2	0	2
	5	6	0	3	4	2
2	1	5	2	8	8	9

3 Only one number has a digit in the millions, so this number is the largest.

2,152,889

4 Two numbers have hundreds of thousands, so are next. Check the number of hundreds of thousands.

In order, these numbers are: 844,202; 560,342

5 What is your answer? Check the order.

2,152,889; 844,202; 560,342

Try this

1 Write these numbers in order, starting with the **largest**.

624,429 62,994 639,422 364,922 349,244

[ ] [ ] [ ] [ ] [ ]

largest **smallest**

Rounding

To achieve the expected standard, you need to:

★ **round** any whole number to the nearest 10; 100; 1,000; 10,000; or 100,000.

What you need to know

Rounding means approximating a number. Numbers can be rounded up or down to different **multiples** of 10, 100, 1,000, etc. When rounding to the nearest 10:

- If the value of the digit is 5 or above, round up.
- If the value of the digit is 4 or below, round down.

Let's practise

Round 52,845 to:

a) the nearest 10 ☐

b) the nearest 100 ☐

c) the nearest 1,000 ☐

1. Read the question and read it again. What is it asking?
   Round the number 52,845 to the nearest 10; 100; and 1,000
2. Start with a). Remember the rules about rounding.
   To round to the nearest 10, I look at the ones digit. It is 5, so I round up to 52,850
3. Now b).
   To round to the nearest 100, I look at the tens digit. It is 4, so I round down to 52,800
4. Now c).
   To round to the nearest 1,000, I look at the hundreds digit. It is 8, so I round up to 53,000
5. What is your answer?
   a) 52,850 b) 52,800 c) 53,000
6. Check your answer.

Try this

1 A number rounds to 8,500 when rounded to the nearest 100. Write two possible numbers.

☐

☐

Top tip

- Whether you are rounding to the nearest 10, 100 or 1,000, look at the digit to the right of the digit you are rounding. If it is 1 to 4, round down; if the digit is 5 or more, round up.

Place value of decimal numbers

To achieve the expected standard, you need to:

★ know the place value of digits in numbers with up to two decimal places.

What you need to know

- The decimal point separates whole numbers on the left from parts of whole numbers on the right.
- The place value of each digit depends on its position after the decimal point: tenths and hundredths.

Let's practise

What is the value of the 7 in each of these numbers?

7.29 ______ 5.78 ______ 8.17 ______

1 Read the question and read it again. What is it asking?

Identify the value of each 7

2 Each time you move one column to the right, the value of the column is ten times smaller.
A **O**ne divided by 10 is a **t**enth (t).
A **t**enth divided by 10 is a **h**undredth (h).
Insert the numbers, using the decimal point as a guide.

O	.	t	h
7	.	2	9
5	.	7	8
8	.	1	7

3 Find the value of the digit 7 in each number by reading the title of its column. What is your answer?

In 7.29, the 7 is worth 7 ones.
In 5.78, the 7 is worth 7 tenths.
In 8.17, the 7 is worth 7 hundredths.

4 Check your answer.

$7, \frac{7}{10}, \frac{7}{100}$

Try this

1 What is the value of the 4 in each of these numbers?

5.74 ______ 8.4 ______ 0.64 ______ 4.95 ______ 15.40 ______

Negative numbers

To achieve the expected standard, you need to:

- ★ use negative numbers in practical contexts
- ★ calculate intervals across zero.

What you need to know

- **Negative numbers** are numbers less than zero.

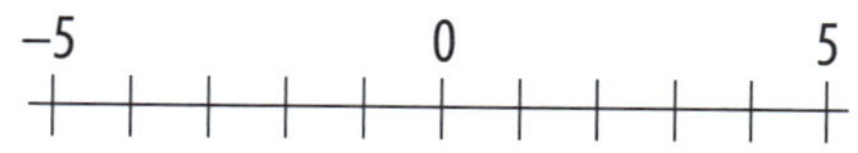

Let's practise

Find the difference between 3 and –5 ☐

1 Read the question and read it again. What is it asking?

Work out the difference between 3 and –5

2 Picture the numbers, and draw a sketch of a number line.

3 is positive and –5 is negative.

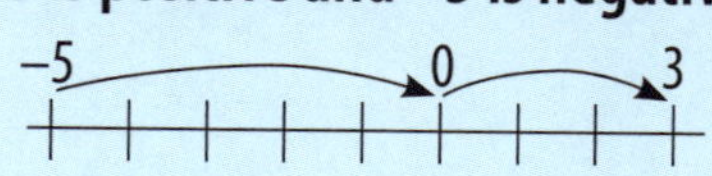

3 Count along the line using 0 as a stopping point.

From –5 to 0 is 5 and from 0 to +3 is 3, so 5 + 3 = 8

4 Check your calculation.

Counting back 8 from 3 through 0, do I reach –5 again? Yes. The difference is 8

Try this

1 Find the difference between –15 and 9 ☐

2 The temperature rises from –8°C to 7°C.
By how many degrees does it rise? ☐ °C

3 The difference between two temperatures is 8°C. One temperature is 6°C. What could the other two temperatures be?

☐ °C ☐ °C

Top tip

- A mistake can be made by counting all the numbers, including the start number. It's the jumps *between* the numbers that you need to count.

Addition

To achieve the expected standard, you need to:

- ★ secure adding whole numbers up to four digits
- ★ add whole numbers with more than four digits
- ★ use formal written methods.

What you need to know

- *Sum, total of, increase by, plus* or *altogether* are all used to mean 'add'.
- When using the column addition method, make sure the numbers are written in the correct column.

Let's practise

Work out the sum of 51,680 and 2,948 ☐

1. Read the question and read it again. What is it asking?

   51,680 + 2,948 = ?

2. Estimate an answer by rounding each number to the nearest 1,000.

   52,000 + 3,000 = 55,000

3. Write the calculation, using place value to line up the digits.

$$\begin{array}{rccccc} & 5 & 1 & 6 & 8 & 0 \\ + & & 2 & 9 & 4 & 8 \\ \hline & & & & & \\ \hline \end{array}$$

4. Work from the ones column to the tens of thousands column, carrying when needed.

$$\begin{array}{rccccc} & 5 & 1 & 6 & 8 & 0 \\ + & & 2 & 9 & 4 & 8 \\ \hline & 5 & 4 & 6 & 2 & 8 \\ \hline & & {\scriptstyle 1} & {\scriptstyle 1} & & \end{array}$$

5. What is your answer? Check your answer is close to your estimate.

   54,628 is close to 55,000. My answer is 54,628

Try this

1 62,341 + 15,420 = ☐

2 A theatre has two tiers. Tier 1 has 1,250 seats. Tier 2 has 1,675 seats.
How many seats are there in total? ☐

3

$$\begin{array}{rcccc} & 6 & 3 & \square & 5 \\ + & & \square & 2 & \square \\ \hline & 7 & 0 & 6 & 8 \\ \hline \end{array}$$

Subtraction

To achieve the expected standard, you need to:

- ★ secure subtracting whole numbers up to four digits
- ★ subtract whole numbers with more than four digits
- ★ use formal written methods.

What you need to know

- The words *subtract, difference between, decrease by, reduce by, take away, minus* or *less* are all used to mean 'subtract'.

Let's practise

Work out the difference between 24,716 and 7,470 ☐

1 Read the question and read it again. What is it asking?

24,716 – 7,470 = ?

2 Estimate an answer by rounding each number to the nearest 1,000.

25,000 – 7,000 = 18,000

3 Write the subtraction, using place value to line up the digits. Start by subtracting the ones. Next subtract the tens. As the bottom number is larger, exchange a hundred from the next column. Continue with the other columns.

6 – 0 = 6, so write 6 in the ones column.

	$^{1}\not{2}$	$^{1}4$	$^{6}\not{7}$	$^{1}1$	6
–		7	4	7	0
	1	7	2	4	6

4 What is your answer? Check your answer is close to your estimate.

17,246 is close to 18,000. My answer is 17,246

Try this

1 16,403 – 8,761 = ☐

2 Lucky Luke wins £480,921 on the lottery.

He buys a car that costs £47,750, and a house that costs £305,000

How much money will he have left? ☐

3

	8	☐	5	9
–	☐	3	4	☐
	2	5	1	3

Multiplying and dividing by 10 and 100

To achieve the expected standard, you need to:

★ multiply and divide whole numbers and decimal numbers with up to two decimal places by 10 or 100.

What you need to know

- When multiplying or dividing by 10 or 100, the digits stay locked together and move left or right.
- Place value will help you when multiplying and dividing with decimal numbers.

Let's practise

3.4 × 100 = ☐ 675 ÷ 10 = ☐

1 Read the question and read it again. What is it asking?

Multiply 3.4 by 100 and divide 675 by 10

2 Do 3.4 × 100 first.
Each time a number moves one column to the left, it gets 10 times larger.
To multiply by 100, move the digits two columns to the left.
You need to add a 0 as a placeholder in the ones column.

1000	100	10	1		$\frac{1}{10}$	$\frac{1}{100}$	$\frac{1}{1000}$
Th	H	T	O	.	t	h	th
			3	.	4		
	3	4	0	.			

3 Now do 675 ÷ 10
Each time a number moves one column to the right, it gets 10 times smaller.
To divide by 10, move the digits one column to the right.
Don't forget to use the decimal point.

1000	100	10	1		$\frac{1}{10}$	$\frac{1}{100}$	$\frac{1}{1000}$
Th	H	T	O	.	t	h	th
	6	7	5	.			
		6	7	.	5		

4 What are your answers? Check they look right.

3.4 × 100 = 340
675 ÷ 10 = 67.5

Try this

1 a) 1,532 × 10 = ☐ b) 1,532 ÷ 100 = ☐

2 a) 63 ÷ ☐ = 6.3 b) 63 × 100 = ☐

3 a) ☐ × 10 = 95 b) ☐ ÷ 100 = 0.77

Multiples and factors

To achieve the expected standard, you need to:

★ recognise and use **multiples** and **factors**.

What you need to know

- A multiple is made by multiplying one number by another.
- A **factor** is a number that divides into another number without leaving a remainder.

Let's practise

List all the factors of 24 and give one multiple of 24

____

1	Read the question and read it again. What is it asking?	Write all the factors of 24, and one multiple of 24
2	What do you know about factors?	The factors of 24 are numbers that go exactly into 24
3	Be systematic. Start with number 1 and check every number up to 24.	$24 \div 1 = 24$, so 1 and 24 are factors $24 \div 2 = 12$, so 2 and 12 are factors $24 \div 3 = 8$, so 3 and 8 are factors $24 \div 4 = 6$, so 4 and 6 are factors
4	Now, find one multiple of 24.	There are lots of multiples of 24: $24 \times 2 = 48$, $24 \times 3 = 72$, and so on.
5	Write down your answer and check it.	1; 2; 3; 4; 6; 8; 12; and 24 are the factors of 24 72 is a multiple of 24

Try this

1 List all the factors of 72

☐ × ☐

☐ × ☐

☐ × ☐

☐ × ☐

☐ × ☐

☐ × ☐

2 Add these numbers to the Venn diagram.

16 21 28 42 50

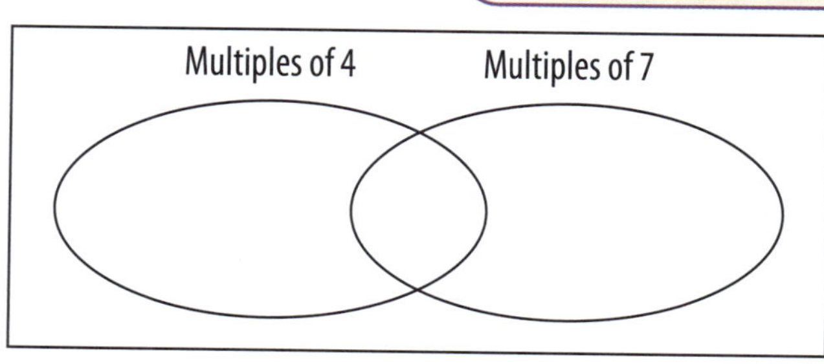

Top tip

- For positive numbers, a factor is never bigger than the number.

Multiplying by larger numbers

To achieve the expected standard, you need to:

★ draw upon multiplication facts up to 12 × 12 and place value to:
  - multiply numbers with up to four digits by a one-digit number using short multiplication
  - multiply numbers with up to four digits by a two-digit number using the formal long multiplication method and become more confident with larger numbers.

What you need to know

- When multiplying by a one-digit number, use the short multiplication method.
- When multiplying by a two-digit number, use the long multiplication method.

Let's practise

1,346 × 7 = ☐

1 Read the question and read it again. What is it asking?

Multiply 1,346 by 7

2 Estimate the answer using rounding to help you.

1,346 is about 1,000 and 7 is about 10
1,000 × 10 = 10,000
The answer will be about 10,000

3 First multiply the digit in the ones column by 7. Put the 2 in the ones column. Carry the 4 tens into the next column so it represents 40.

$$\begin{array}{rcccc} & 1 & 3 & 4 & 6 \\ \times & & & & 7 \\ \hline & & & & 2 \\ & & & {\scriptstyle 4} & \end{array}$$

$6 \times 7 = 42$

4 Next multiply the digit in the tens column by 7. Remember to add the extra 4 tens that have been carried over, so 28 + 4 = 32. Carry the 3 tens into the next column so it represents 300.

$$\begin{array}{rcccc} & 1 & 3 & 4 & 6 \\ \times & & & & 7 \\ \hline & & & 2 & 2 \\ & & {\scriptstyle 3} & {\scriptstyle 4} & \end{array}$$

$4 \times 7 = 28$

5 Continue by multiplying both the digit in the hundreds column and then the digit in the thousands column by 7. Remember to carry over the tens part of any answer.

$$\begin{array}{rcccc} & 1 & 3 & 4 & 6 \\ \times & & & & 7 \\ \hline & 9 & 4 & 2 & 2 \\ & {\scriptstyle 2} & {\scriptstyle 3} & {\scriptstyle 4} & \end{array}$$

6 What is your answer? Check it is close to your estimate.

9,422 is close to 10,000
My answer is 9,422

Let's practise

267 × 68 = ☐

1 Read the question and read it again. What is it asking?

Multiply 267 by 68

2 Estimate the answer.

267 is about 300 and 68 is about 70
$300 \times 70 = 21{,}000$
The answer will be about 21,000

3 Do the multiplication in two parts. First multiply by the ones so start with multiplying by 8.

$$\begin{array}{rcccc} & & 2 & 6 & 7 \\ \times & & & 6 & 8 \\ \hline & 2 & {}_5 1 & {}_5 3 & 6 \end{array}$$

$267 \times 8 = 2{,}136$

4 Next, multiply by the tens. Place a zero in the ones column to represent multiplying by 10, and then multiply by the 6.

$$\begin{array}{rcccc} & & 2 & 6 & 7 \\ \times & & & 6 & 8 \\ \hline & 2 & {}_5 1 & {}_5 3 & 6 \\ 1 & {}_4 6 & {}_4 0 & 2 & 0 \\ \hline \end{array}$$

$267 \times 6 = 1{,}602$
$267 \times 60 = 16{,}020$

5 Add the two together. What is your answer? Check it is close to your estimate.

$$\begin{array}{rcccc} & & 2 & 6 & 7 \\ \times & & & 6 & 8 \\ \hline & 2 & 1 & 3 & 6 \\ 1 & 6 & 0 & 2 & 0 \\ \hline 1 & 8 & 1 & 5 & 6 \end{array}$$

18,156 is close to 21,000
My answer is 18,156

Try this

1 568 × 9 = ☐

2 2,138 × 6 = ☐

3 436 × 24 = ☐

4 5,427 × 36 = ☐

5 In a supermarket storeroom, there are:

- 7 boxes of mango juice cartons
- 5 boxes of peach juice cartons
- 13 boxes of orange juice cartons.

There are **214 cartons** in every **box**.

How many cartons of juice are there **altogether**? ☐ cartons

Top tips

- Don't forget the placeholder 0 when multiplying by the tens digit.
- Remember to add on any numbers that you have carried over to the next column.

Square numbers

To achieve the expected standard, you need to:

★ recognise and use **square numbers** up to 144.

What you need to know

- A **square number** is the number made by multiplying a number by itself (e.g. $3 \times 3 = 9$).
- 3×3 can be written as 3^2

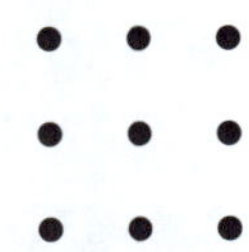

Let's practise

Find two square numbers that add up to 100 ☐ ☐

1	Read the question and read it again. What is it asking?	Find two square numbers that add up to 100
2	List all the square numbers up to 100.	1; 4; 9; 16; 25; 36; 49; 64; 81; 100
3	Look for two square numbers that add up to 100 It would be helpful to look at the ones digits to find a pair that add to 10.	$36 + 64 = 100$
4	What is your answer? Check it looks right.	36 and 64

Try this

1 a) 4^2 ☐ b) 9^2 ☐

2 Find **three** square numbers that add up to 50 ☐ ☐ ☐

3 A square has an area of 144 cm^2.
What is the length of each side? ☐ cm

4 $8^2 + \square^2 = 100$

Top tip

- Learn the squares of 1 to 12. These are: 1; 4; 9; 16; 25; 36; 49; 64; 81; 100; 121; 144.

Short division

To achieve the expected standard, you need to:

- ★ use the formal written method of short division to divide numbers with up to four digits by a one-digit number
- ★ divide whole numbers mentally, drawing upon multiplication facts and place value, and begin to use these facts to work with larger numbers.

What you need to know

- You can check your answer to the division calculation using multiplication because it is the **inverse**.

Let's practise

$2{,}780 \div 4 = \square$

1	Read the question and read it again. What is it asking?	Divide 2,780 by 4
2	Set out the calculation, and start by looking at the first digit of 2,780 $2 \div 4 = 0$ remainder 2 Put the 2 next to the 7.	$4 \overline{\smash{)}\,2 \ \ {}^{2}7 \ \ 8 \ \ 0}$
3	$27 \div 4 = 6$ remainder 3 Put the 6 in the answer. Put the 3 next to the 8.	$\begin{array}{r} 6 \ \ \ \ \ \ \\ 4 \overline{\smash{)}\,2 \ \ {}^{2}7 \ \ {}^{3}8 \ \ 0} \end{array}$
4	$38 \div 4 = 9$ remainder 2 Put the 9 in the answer. Put the 2 next to the 0.	$\begin{array}{r} 6 \ \ 9 \ \ \ \\ 4 \overline{\smash{)}\,2 \ \ {}^{2}7 \ \ {}^{3}8 \ \ {}^{2}0} \end{array}$
5	$20 \div 4 = 5$ Put the 5 in the answer.	$\begin{array}{r} 6 \ \ 9 \ \ 5 \\ 4 \overline{\smash{)}\,2 \ \ {}^{2}7 \ \ {}^{3}8 \ \ {}^{2}0} \end{array}$
6	Check your answer by doing a multiplication. When you are sure it is correct, write your answer in the box.	I can check by working out that: $695 \times 4 = 2{,}780$

Try this

1. $721 \div 7 = \square$
2. $5{,}216 \div 8 = \square$
3. $7{,}893 \div 8 = \square$
4. $4{,}360 \div 20 = \square$

Long division

To achieve the expected standard, you need to:

★ become more confident using long division to divide numbers with up to four digits by a two-digit whole number.

What you need to know

- When you divide by a one-digit number, you might do the calculation in your head or use short division. Dividing by a two-digit number is harder, and it's best to use a **long division** layout.

```
    5 2
7 | 3 6 ¹4
```

Let's practise

$2{,}826 \div 12 = \square$

1 Read the question and read it again. What is it asking?

Divide 2,826 by 12

2 Set out the calculation. Start by looking at the first two digits of 2,826.

```
12 | 2 8 2 6
```

3 Divide 28 by 12 to find how many times 12 goes into 28. Write 2 above the line in the answer space and record 24 (12 × 2) below 28. Subtract 24 from 28 to find the remainder: 4. Bring down the digit 2.

```
       2
12 | 2 8 2 6
     2 4 ↓
       4 2
```

4 Divide 42 by 12 to find how many times 12 goes into 42.
Write 3 above the line in the answer space and record 36 (12 × 3) below 42.
Subtract 36 from 42 to find the remainder, 6.
Bring down the digit 6 and continue by dividing 66 by 12.
66 ÷ 12 = 5 r6

```
       2 3 5 r6
12 | 2 8 2 6
     2 4
       4 2
       3 6 ↓
         6 6
         6 0
           6
```

5 What is your answer? Don't forget the remainder.

For your answer, you can use a remainder (235 r6), a fraction ($235\frac{6}{12}$ or $235\frac{1}{2}$) or a decimal (235.5).

Try this

1 $351 \div 13 = \square$ **2** $1{,}710 \div 18 = \square$ **3** $5{,}168 \div 34 = \square$

Prime numbers

To achieve the expected standard, you need to:

★ recognise and use **prime numbers** less than 20.

What you need to know

- A **prime number** has only *two* factors: 1 and itself. It cannot be divided by another number to give a whole number. 3; 5; and 7 are all prime numbers.

Let's practise

Circle all the numbers that are prime numbers.

1	2	3	4	5	6	7	8	9	10
11	12	13	14	15	16	17	18	19	20

1. Read the question and read it again. What is it asking?

   Circle all the prime numbers.

2. 2 is a prime number because it only has two factors, 1 and itself. All other multiples of 2 (all even numbers) will have 2 as a factor and so cannot be prime.

   So I can cross out 4; 6; 8; 10; 12; 14; 16; 18; and 20

3. 3 and 5 are also prime, but other multiples of 3 and 5 are not prime, so they can be crossed out.

   I can cross out 9 and 15

4. What does that leave? What is your answer?

   2; 3; 5; 7; 11; 13; 17; and 19 are left and should be circled.

Try this

1 Write the next four prime numbers after 10 ☐ ☐ ☐ ☐

2 Kate says 2 is not a prime number because it is even. Is she correct? YES / NO

Explain how you know. ______________________________

Top tip

- Prime numbers must have only two factors, so 1 is not a prime number.

Fractions of amounts

To achieve the expected standard, you need to:
- ★ calculate simple fractions of whole numbers and quantities.

What you need to know

- A unit **fraction** has 1 as the **numerator** (e.g. $\frac{1}{3}$, $\frac{1}{8}$, $\frac{1}{9}$).
- $\frac{\text{Numerator}}{\text{Denominator}}$
- Finding fractions of a number is linked to division (e.g. $\frac{1}{7}$ of 56 has the same answer as 56 divided by 7).

Let's practise

$\frac{3}{4}$ of 36 = ☐

1 Read the question and read it again. What is it asking?	Find three-quarters of 36
2 First, find $\frac{1}{4}$ of 36.	$\frac{1}{4}$ of $36 = 9$ because $36 \div 4 = 9$
3 Now focus on the numerator, 3. Multiply $\frac{1}{4}$ of 36 by 3 to find $\frac{3}{4}$ of 36.	$\frac{3}{4}$ of $36 = 3 \times \frac{1}{4}$ of $36 = 3 \times 9 = 27$
4 What is your answer?	$\frac{3}{4}$ of $36 = 27$
5 Check your answer using the reverse order. Divide 27 first by 3 to give the value of $\frac{1}{4}$ and then multiply by 4 to give the value of the whole. You should return to the original number, 36.	$27 \div 3 = 9$ $9 \times 4 = 36$

Top tips

- Do one line of working at a time.
- Check your answer by working it out in reverse order.

Try this

1 Find these values:

a) $\frac{1}{2}$ of 36 ☐ b) $\frac{2}{3}$ of 24 ☐ c) $\frac{1}{4} \times 52$ ☐ d) $\frac{3}{5}$ of 60 ☐

2 a) $\frac{3}{4}$ of ☐ $= 45$ b) $\frac{3}{5}$ of ☐ $= 24$

Mixed numbers

To achieve the expected standard, you need to:

★ recognise **mixed numbers** and **improper fractions** and convert from one to the other.

What you need to know

- A **mixed number**, such as $3\frac{1}{4}$, has a whole number part (3) and a fractional part ($\frac{1}{4}$) and it means $3 + \frac{1}{4}$.
- **Improper fractions** are fractions where the numerator is larger than the denominator. Improper fractions can be turned into mixed numbers: $\frac{11}{4}$ can be written as $2\frac{3}{4}$.

Let's practise

Convert $3\frac{1}{4}$ to an improper fraction. ☐

1. Read the question and read it again. What is it asking?

   Convert the mixed number three and a quarter into an improper fraction.

2. What denominator will the improper fraction need?

   $\frac{1}{4}$ **has 4 as the denominator.**

3. Convert the whole number 3 into an improper fraction. Remember, each whole number will have four quarters, so three whole numbers will have 3×4 quarters. Don't forget to add the extra $\frac{1}{4}$.

   $3 \times \frac{4}{4} = \frac{12}{4}$

   $\frac{12}{4} + \frac{1}{4} = \frac{13}{4}$

4. What is your answer? Check your working.

   $\frac{13}{4}$

Try this

1 Convert these mixed numbers to improper fractions.

a) $2\frac{1}{2}$ ☐ b) $5\frac{1}{10}$ ☐ c) $3\frac{2}{5}$ ☐

2 Circle **all** the fractions that are equivalent to $2\frac{2}{3}$

$\frac{7}{5}$ $\frac{15}{4}$ $\frac{16}{6}$ $\frac{11}{6}$ $\frac{24}{9}$ $\frac{20}{9}$

Top tip

- Remember that the denominator will stay the same when you change mixed numbers into improper fractions or improper fractions into mixed numbers.

Equivalent fractions

To achieve the expected standard, you need to:

- ★ recognise and use **equivalent fractions**
- ★ find equivalent fractions with lower denominators
- ★ rewrite a pair of fractions so they share the same denominator.

What you need to know

- **Equivalent fractions** have the same value even though they look different.
- For each fraction, there is a family of equivalent fractions: $\frac{2}{5}$ $\frac{4}{10}$ $\frac{6}{15}$ $\frac{8}{20}$

Let's practise

$$\frac{2}{3} = \frac{\square}{12}$$

1 Read the question and read it again. What is it asking?	Find the equivalent fraction to $\frac{2}{3}$, with 12 as the denominator.
2 To find an equivalent fraction, multiply the numerator and denominator by the same number. Look at the denominators. What has 3 been multiplied by to give 12?	$\frac{2}{3} = \frac{\boxed{?}}{12}$ $3 \times 4 = 12$
3 To keep the numerator and denominator in proportion, multiply the numerator by 4 as well.	$2 \times 4 = 8$ So, $\frac{2}{3} = \frac{8}{12}$
4 What is your answer? Check your working.	$\frac{2}{3} = \frac{\boxed{8}}{12}$

Try this

1 a) $\frac{1}{5} = \frac{\square}{10}$ b) $\frac{5}{9} = \frac{15}{\square}$ c) $\frac{2}{3} = \frac{6}{\square}$ or $\frac{\square}{21}$

2 Simplify:

a) $\frac{10}{12} = \square$ b) $\frac{12}{15} = \square$ c) $\frac{20}{28} = \square$ d) $\frac{400}{800} = \square$

3 Change $\frac{3}{4}$ and $\frac{7}{10}$ to equivalent fractions that have the same denominator. $\square$ $\square$

Top tip

- Always remember to multiply or divide the numerator and denominator by the same number.

Adding and subtracting fractions

To achieve the expected standard, you need to:

- ★ add and subtract fractions with the same denominator, using mixed numbers where appropriate for the context
- ★ add and subtract fractions with denominators that are multiples of the same number, and become more confident with more complex calculations.

What you need to know

- Denominators in fractions need to be the same before they can be added or subtracted. This can be done by using equivalent fractions (see page 25).

Let's practise

$$7\frac{5}{8} - 3\frac{1}{4} = \square$$

1 Read the question and read it again. What is it asking?	Subtract three and one-quarter from seven and five-eighths.
2 The denominators are not the same. The fractions must be changed so they have the same denominator. You can change quarters to eighths.	$\frac{1}{4}$ is equivalent to $\frac{2}{8}$
3 Rewrite the question and then do the subtraction.	$7\frac{5}{8} - 3\frac{2}{8} =$
4 Subtract the whole numbers. Then subtract the fractions.	$7 - 3 = 4$ $\quad \frac{5}{8} - \frac{2}{8} = \frac{3}{8}$
5 What is your answer? Check your working.	$4\frac{3}{8}$

Try this

1 a) $\frac{6}{7} - \frac{2}{7} = \square$ b) $\frac{5}{6} - \square = \frac{3}{6}$ c) $\frac{3}{4} + \frac{1}{8} = \square$

2 a) $1\frac{5}{8} - \frac{2}{8} = \square$ b) $2\frac{3}{10} + 2\frac{4}{10} = \square$

Top tip

- It is often useful to use improper fractions for subtraction calculations that cross a whole number boundary, e.g. $1\frac{1}{5} - \frac{2}{5} = \frac{6}{5} - \frac{2}{5} = \frac{4}{5}$.

Fractions and their decimal equivalents

To achieve the expected standard, you need to:

- ★ recognise and write decimal equivalents for $\frac{1}{4}$, $\frac{1}{2}$, $\frac{3}{4}$ and any number of fifths, tenths or hundredths and read and write decimals as fractions
- ★ recognise and use thousandths and relate them to tenths, hundredths and decimal equivalents.

What you need to know

- Fractions can be written as decimal numbers using the decimal point to show where the fractional part starts.

Fraction		as tenths or hundredths		ones		$\frac{?}{10}$	$\frac{?}{100}$		
$\frac{4}{5}$	=	$\frac{8}{10}$	=	0	.	8		=	0.8
$\frac{3}{4}$	=	$\frac{75}{100}$	=	0	.	7	5	=	0.75

Let's practise

Write the decimal equivalent of $\frac{1}{4}$ ☐

1 Read the question and read it again. What is it asking?

Turn one-quarter into a decimal fraction.

2 Remember the line in a fraction is a dividing line, so $\frac{1}{4}$ also means $1 \div 4$.

$$4 \overline{)\,1\,.\,{}^{1}0\;{}^{2}0} = 0\,.\,2\,5$$

3 What is your answer? Check your working.

$\frac{1}{4} = 0.25$

Try this

1 Write the missing fraction or decimal.

a) $\frac{6}{100}$ = ☐   b) 0.55 = ☐   c) $\frac{3}{5}$ = ☐

2 Write each decimal number to its equivalent fraction.

a) $\frac{1}{4}$ ☐   b) $\frac{2}{5}$ ☐   c) $\frac{3}{4}$ ☐   d) $\frac{2}{10}$ ☐

Adding and subtracting decimals

To achieve the expected standard, you need to:

★ add and subtract decimal numbers that have the same number of decimal places.

What you need to know

- The decimal point separates whole numbers on the left of the decimal point from parts of whole numbers on the right.
- When adding or subtracting decimal numbers, line up decimal points and check that the tens, ones, tenths and hundredths are in the correct place value column.

Let's practise

78.53 – 6.35 = ☐

1 Read the question and read it again. What is it asking?	Subtract 6.35 from 78.53
2 Estimate the answer.	Rounding to the nearest 10: 78.53 is near 80 6.35 is near 10 80 – 10 = 70 So, the answer will be around 70
3 Set out the calculation, lining up the decimal points.	$\begin{array}{rrcrr} & 7 & 8 & . & {}^{4}\not{5} & {}^{1}3 \\ - & & 6 & . & 3 & 5 \\ \hline & 7 & 2 & . & 1 & 8 \end{array}$
4 What is your answer? Don't forget to write the decimal point.	72.18
5 Check your answer using the inverse operation of addition.	72.18 + 6.35 = 78.53

Try this

1 Greg and Niv are doing face painting at the school fayre. Greg earns £42.75 and Niv earns £48.78

They give £30.50 to charity and keep the rest. How much money do they keep? ☐

Top tip

- Always check your answers. Check addition by subtracting. Check subtraction by adding.

Multiplying decimals

To achieve the expected standard, you need to:

★ multiply a one-digit decimal number by a one-digit number.

What you need to know

- Multiplication facts are essential when multiplying decimals.
- Place value is important when multiplying decimal numbers.

Let's practise

$0.7 \times 9 = \square$

1	Read the question and read it again. What is it asking?	**Multiply 0.7 by 9**
2	Picture the numbers and estimate the answer.	**0.7 rounds to 1, so the answer must be less than $1 \times 9 = 9$**
3	Use known facts.	**$7 \times 9 = 63$** **0.7 is 10 times smaller than 7** **0.7×9 will be 10 times smaller than 7×9** **So, $63 \div 10 = 6.3$**
4	What is your answer?	**6.3**
5	Check your answer against your estimate.	**6.3 is less than 9**

Try this

1. $0.5 \times 7 = \square$
2. $0.6 \times 8 = \square$
3. $0.09 \times 4 = \square$
4. $0.3 \times 4 = \square$
5. $\square \times 8 = 3.2$
6. $0.07 \times \square = 0.42$
7. $\square \times 5 = 0.45$
8. $0.3 \times \square = 2.7$

Top tip

- Use your known facts when multiplying decimal numbers. If $3 \times 7 = 21$, then 3×0.7 will be 10 times smaller than 21 and so the answer is 2.1

Percentages as fractions and decimals

To achieve the expected standard, you need to:

- ★ recognise and use the equivalences between simple fractions, decimals and percentages
- ★ become more confident with calculating other decimal fraction equivalents.

What you need to know

- The **percentage** symbol (%) means 'out of 100'.
- Fractions with 100 as the denominator can be written as decimals and percentages (e.g. $\frac{74}{100} = 0.74 = 74\%$).
- Other fractions can be written as decimals by dividing the numerator by the denominator (e.g. $\frac{1}{4} = 1 \div 4 = 0.25 = 25\%$).

Let's practise

Write 35% as a fraction and as a decimal. ☐ = ☐

1. Read the question and read it again. What is it asking?
   35% is a percentage. I have to write it as a fraction and as a decimal.
2. Remember what % means. Find the fraction equivalent.
   35% means 35 out of 100
   $35\% = \frac{35}{100}$
3. Remember that fractions with 100 as the denominator are easy to write as decimals.
   $\frac{35}{100} = 35 \div 100 = 0.35$
4. What is your answer? Check your working.
   $35\% = \frac{35}{100} = 0.35$

Try this

1 Write 75% as a fraction and as a decimal.

2 Write 0.6 as a fraction and a percentage.

3 Write $\frac{3}{10}$ as a decimal and a percentage.

Top tip

- Percentages can be written as fractions over 100.

Finding percentages

To achieve the expected standard, you need to:

★ find simple percentages of whole numbers and quantities.

What you need to know

- Percentages can be written as fractions over 100: 25% = $\frac{25}{100}$.
- Percentages can be written as decimals: 25% = 0.25.
- 10% is equal to $\frac{10}{100}$ and can be simplified to $\frac{1}{10}$ by dividing both the numerator and denominator by 10.

Let's practise

Find 30% of 60 ☐

1	Read the question and read it again. What is it asking?	Calculate 30% of 60
2	30% is equivalent to 10% × 3 so start by finding 10% of 60. Remember that finding 10% is the same as finding $\frac{1}{10}$ or dividing by 10.	$30\% = 10\% + 10\% + 10\%$ or $10\% \times 3$ 10% of $60 = 60 \div 10 = 6$ $30\% = 6 + 6 + 6$ or 6×3 $= 18$
3	What is your answer? Check your working using a reverse calculation.	18 $18 \div 3 = 6$, which is 10% $6 \times 10 = 60$, which is 100%

Try this

1 Find 20% of 80 ☐

2 Find 25% of 120 ☐

3 Find 30% of 300 ☐

4 Find 40% of 250 ☐

5 Janice makes 250 ml of a fruit smoothie.

20% of the smoothie is honey, 25% is bananas and the rest is vanilla yoghurt.

How many millilitres of vanilla yoghurt does she use? ☐ ml

Top tip

- There are different ways to find percentages of numbers. Use the method you feel most comfortable with.

Ratio and proportion

To achieve the expected standard, you need to:

- ★ use simple **ratio** to compare quantities
- ★ estimate the distance on a map using a simple scale.

What you need to know

- **Ratio** is the relationship between two or more quantities. It compares 'part with part' (e.g. the ratio 3:1 means 3 for every 1).
- **Proportion** is the relationship between part of something and the whole thing. It compares 'part with whole' (e.g. 1 out of 4 is the same as 2 out of 8). Think of proportion as a **fraction** ($\frac{1}{4} = \frac{2}{8}$). We can also write this proportion as 0.25 or 25%.

Let's practise

A necklace is made using 2 blue beads for every 3 red beads. There are 35 beads altogether. How many red beads are used? ☐

1. Read the question and read it again. What is it asking?

   Work out the number of red beads in the necklace using proportion.

2. Picture the necklace.

   The pattern is repeated every 5 beads.

3. So how many patterns will there be in the complete necklace, using 35 beads?

   $35 \div 5 = 7$
   There will be 7 patterns.

4. How many red beads will there be?

   7 (patterns) × 3 (beads) = 21

5. How can you check your answer?

   By counting the blue beads too.
   $7 \times 2 = 14$, $14 + 21 = 35$

Try this

1 Jaime baked some cupcakes.

For every 6 cupcakes Jaime baked, only 4 cupcakes were iced.

Altogether, 20 cakes were iced.

How many cakes did Jaime bake? ☐ cakes

2 A drink is made using water and orange squash in the ratio 5:1

How much water will be added to 50 ml of orange squash? ☐ ml

Top tip

- Don't forget to include units in your answer.

Algebra

To achieve the expected standard, you need to:
★ use simple formulae expressed in words.

What you need to know

- In algebra, words or letters are used to represent the amounts in a problem, instead of actual numbers.

Let's practise

This formula gives the exact time to cook a turkey:
$T = 40m + 20$
In the formula:
- T = the cooking time
- m = the mass of the turkey in kilograms.

How long should it take to cook a turkey with a mass of 3.5 kg?

[ ] minutes

1	Read the question and read it again. What is it asking?	Calculate the time it takes to cook a turkey with a mass of 3.5 kg using the formula.
2	Check the formula.	$T = 40m + 20$ T is the answer to the question, which is the time to cook the turkey.
3	You know the mass of the turkey is 3.5 kg so insert 3.5 for m (the mass of the turkey) and calculate the answer.	$T = 40 \times 3.5 + 20$ $T = 140 + 20$ $= 160$
4	What is your answer? Check your working.	160 minutes

Try this

1 Using the same formula, how long should it take to cook a turkey with a mass of 5.25 kg?

[ ] minutes

2 Using the same formula, a turkey takes two hours to cook. How heavy is it? [ ] kg

Top tip

- When you know the values of letters (unknowns) in a formula, just substitute the values for the letters and complete the calculation.

Sequences

To achieve the expected standard, you need to:

- ★ count forwards or backwards in steps of any whole number with one significant figure
- ★ generate, describe and complete linear number **sequences**.

What you need to know

- A **sequence** is a set of numbers that follow a pattern according to a rule.

Let's practise

Continue this sequence and explain the rule in writing.

15 18 21 ☐ ☐ Rule is ____________

1 Read the question and read it again. What is it asking?	Put numbers in the two boxes to continue the sequence 15; 18; 21
2 Picture the numbers.	They start at 15 and get bigger.
3 To find the pattern, check the difference between pairs of numbers that are next to each other by counting on.	$18 - 15 = 3$; $21 - 18 = 3$ So, the difference is 3
4 The difference tells you the rule for the pattern. Use it to work out the missing numbers.	The rule is 'add 3'. $21 + 3 = 24$ $24 + 3 = 27$
5 What is your answer? Check you have stated the rule.	24 and 27 Rule is 'add 3'.

Try this

1 Complete these sequences. For each one, find the missing numbers and state the rule.

a) 23 32 41 ☐ ☐ Rule is ____________

b) 9 5 ☐ ☐ −7 Rule is ____________

c) 50 ☐ ☐ 170 Rule is ____________

2 The rule is add 3,000. What are the missing numbers?

☐ ☐ ☐ 10,500 13,500

Top tip

- If you can't work out the rule, try some numbers to see if they work.

Solving equations

To achieve the expected standard, you need to:

★ find possible values in missing-number problems and equations involving one or two unknowns.

What you need to know

- Algebra is useful for solving problems when you can write the facts as equations.

Let's practise

$100 - \square = 20 \times \square$

1	Read the question and read it again. What is it asking?	Find two numbers: one that is subtracted from 100 and one that multiplies by 20. Remember that $100 - \square$ must have the same value as $20 \times \square$.
2	There are an infinite number of answers to this question, so just try a number.	$100 - 40 = 60$
3	If the first unknown number is 40, the left-hand side must be worth 60 and so must the right-hand side. So the right-hand side must be 20×3.	$60 = 20 \times \square$ $60 \div 20 = 3$
4	An answer could be:	$100 - 40 = 20 \times 3$
5	Try finding some of the other possible values for this equation.	Here are some more possible solutions: $100 - 0 = 20 \times 5$ $100 - 80 = 20 \times 1$

Try this

1 Find a possible solution for each of these equations.

a) $40 + \square = 90 - \square$ b) $10 \times \square = 200 - \square$

2 Each shape stands for a number. Work out the value of each shape.

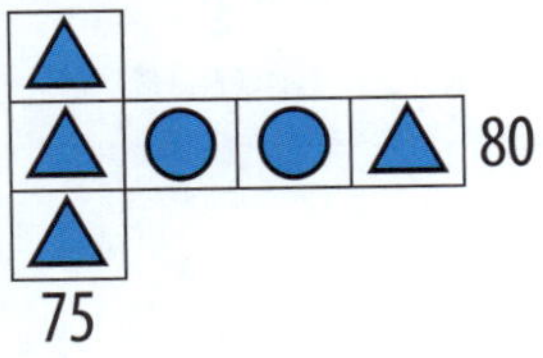

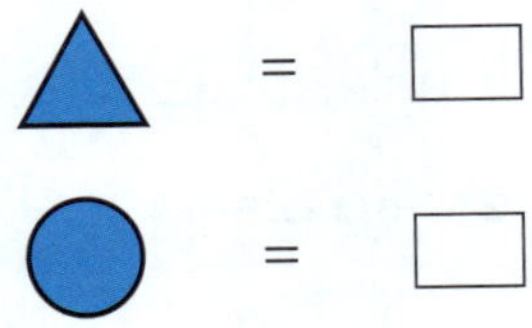

Length

To achieve the expected standard, you need to:

★ use, read, write and convert between metric units of length
★ use all four operations to solve problems involving length
★ use decimals to two decimal places.

What you need to know

- *Kilo* stands for 1,000, *milli* stands for $\frac{1}{1{,}000}$ and *centi* stands for $\frac{1}{100}$.
- **Length** is measured in millimetres, centimetres, metres and kilometres.
- 1 km = 1,000 m, 1 m = 100 cm, 1 cm = 10 mm.

Let's practise

Circle the **shortest** length.

1.5 m   325 cm   2,275 mm

1 Read the question and read it again. What is it asking?

Which is the shortest length?

2 You cannot compare the numbers until all the units are the same. Here we are going to use cm.

1.5 m	325 cm	2,275 mm
× 100		÷ 10
150 cm	325 cm	227.5 cm

3 Now you can compare and find the shortest.

150 cm < 227.5 cm < 325 cm

4 What is your answer? Double check it.

1.5 m is the shortest length.

Try this

1 Circle the **longest** length.

2 m   1,000 mm   150 cm

2 Ben cuts 5 m of wood into 3 planks.

The length of the first plank is 1.67 m.

The length of the second plank is 2.36 m.

Work out the length of the third plank. [   ] m

Top tips

- Make sure every length is in the same unit.
- Know your abbreviations:
  mm = millimetre
  cm = centimetre
  m = metre
  km = kilometre

Mass

To achieve the expected standard, you need to:

- ★ use, read, write and convert between units of **mass**
- ★ use all four operations to solve problems involving mass
- ★ use decimals to two decimal places.

What you need to know

- **Mass** is measured in grams or kilograms.
- *Kilo* stands for 1,000.
- 1 kg = 1,000 g

Let's practise

Which is heavier, 3,500 g or 4 kg? ☐

Explain how you know. ______________________________

____

1	Read the question and read it again. What is it asking?	Decide which mass is heavier.
2	Look at the numbers.	3,500 > 4 but the units are not the same so I cannot make a decision just looking at the numbers.
3	What do you know about grams and kilograms?	1 kg = 1,000 g So I am comparing 3,500 g with 4,000 g and 4,000 g is the heavier.
4	What is your answer? Check you have written an explanation.	4 kg is heavier because 4 kg = 4,000 g and 4,000 g > 3,500 g.

Try this

1 Circle the **heaviest** mass. 250 g 2.5 kg 2 kg

2 Dan and Sam each have a parcel.

Dan's parcel weighs $3\frac{1}{4}$ kg.

Sam's parcel weighs 2.5 kg.

How many more grams does Dan's parcel weigh than Sam's parcel? ☐ g

Top tips

- Make sure every mass is in the same unit before comparing them, or doing addition or subtraction.
- Know your abbreviations: g = gram, kg = kilogram.

Capacity

To achieve the expected standard, you need to:

- ★ use, read, write and convert between metric units of **capacity**
- ★ use all four operations to solve problems involving capacity
- ★ use decimals to two decimal places.

What you need to know

- The **capacity** is the amount that can be contained in a space. For liquids, it is measured in litres or millilitres.
- *Milli* stands for $\frac{1}{1,000}$.
- 1 litre (l) = 1,000 millilitres (ml)

Let's practise

Jack pours a fifth of a litre of squash into a jug.
He adds 700 ml of water.
How much drink is in the jug? [ ]

1	Read the question and read it again. What is it asking?	I need to add together $\frac{1}{5}$ of a litre and 700 ml.
2	Look at the numbers and the units.	The amount of water is in millilitres. The amount of squash is in litres. $\frac{1}{5}$ of a litre is 1,000 ml ÷ 5, which is 200 ml.
3	Now the units are the same, they can be added.	200 ml + 700 ml = 900 ml
4	What is your answer? Check you have included the units.	900 ml

Try this

1 A tank holds 50 litres of water.
How many millilitres is 50 litres? [ ] ml

2 How many 200 ml cups can be filled from a 2.5 litre bottle? [ ] cups

3 Jack fills five 125 ml cups from a full 1 litre bottle.
How much water is left in his bottle? [ ] ml

Top tip

- Take care when reading unlabelled marks on scales. Make sure you know the value of each interval on the scale.

Money

To achieve the expected standard, you need to:

★ use all four operations to solve problems involving money, using decimal places.

What you need to know

- A penny is one hundredth of a pound.
- Decimals are used to record money as pounds and pence.

£5.42

pound sign — £; number of pounds — 5; number of pence — 42; decimal point separating pounds and pence

Let's practise

Gita buys $1\frac{1}{2}$ kilograms of potatoes at £1.34 per kilogram. How much do they cost? ☐

1 Read the question and read it again. What is it asking?	**Find the cost of $1\frac{1}{2}$ kg of potatoes when the price is £1.34 per kilogram.**
2 How much is 1 kilogram of potatoes?	**£1.34**
3 How much is $\frac{1}{2}$ a kilogram of potatoes?	**$\frac{1}{2}$ of £1.34 = £0.67**
4 How much is $1\frac{1}{2}$ kilograms of potatoes?	**£1.34 + £0.67 = £2.01**
5 What is your answer? Check your decimal point is in the right place.	**£2.01**

Try this

1. Henry buys $2\frac{1}{2}$ kilograms of carrots at 90p per kilogram. How much do they cost? ☐
2. Ali buys a magazine for £4.75 and two pens for £1.45 each. He pays with a £20 note. How much change does he get? ☐
3. Jack buys apples costing £2.50 per kilogram. He pays £1.00 How many grams of apples does he buy? ☐ g
4. Gary saves £12.75. Nancy saves double the amount Gary saves. Mandy saves £8.50 more than Nancy. How much do they save altogether? ☐

Top tip

- When writing money amounts, always write two digits after the decimal point. So £3 and 5p is written £3.05, not £3.5 or £3.05p; £7 and 40p is written £7.40, not £7.4 or £7.40p.

Time

To achieve the expected standard, you need to:
- ★ read, write and convert time between analogue (including clock faces using Roman numerals) and digital 12- and 24-hour clocks
- ★ use a.m. and p.m. where necessary.

What you need to know

- 'Quarter to' means the same as '45 minutes past'.
- The time before 12 noon (midday) is a.m. The time from noon until midnight is p.m.
- Analogue time is shown on a clock face with an hour and a minute hand. Digital time uses numbers only.

Let's practise

A clock showing 20 minutes to 5 in the afternoon is 40 minutes slow.

Write the real time using 24-hour notation. ☐

1 Read the question and read it again. What is it asking?

I have to add 40 minutes to find the real time and then write the time using 24-hour notation.

2 Decide the time shown by the analogue clock. Change the time to digital time.

The clock shows 20 minutes to 5. This is 40 minutes past 4, so in digital time it is 4:40

3 The clock is slow by 40 minutes, so add 40 minutes to the clock's time to show the actual time. Do this by counting on.

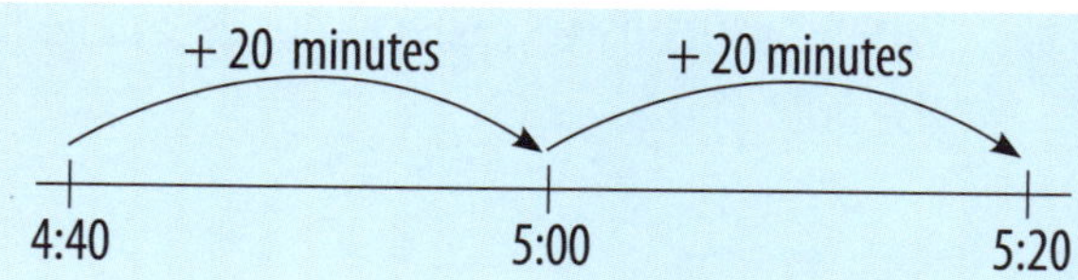

4 Write 5:20 p.m. in 24-hour time. This must use four digits and show the number of hours since midnight.

12 + 5 = 17 so 5 p.m. is 17:00
5:20 p.m. is 17:20

5 What is your answer? Double check it.

17:20

Try this

1 A clock shows 13:25 and is $\frac{3}{4}$ hour slow. Write the actual time in 12-hour time. ☐

2 Write quarter to 8 in the evening using 24-hour digital time. ☐

3 Find the number of minutes between these two times: 16:35 + ☐ minutes = 17:25

Time problems

To achieve the expected standard, you need to:

★ calculate the duration of an event using appropriate units of time.

What you need to know

- To calculate how long an event lasts, you have to calculate a **time interval**.
- A time interval is the difference between two times.
- There are 60 minutes in an hour.

Let's practise

A TV programme starts at 6:45 p.m. and ends at 7:25 p.m. How long is the programme? [ ] minutes

1	Read the question and read it again. What is it asking?	**Work out the difference in time between 6:45 p.m. and 7:25 p.m.**
2	How will you find the difference between these two times? Sketching a number line helps when calculating time intervals.	**From 6:45 to 7 o'clock, it's 15 minutes because there are 60 minutes in the hour. Then it's another 25 minutes to 7:25 p.m.**
3	What calculation do you have to do?	**15 minutes + 25 minutes = 40 minutes**
4	What is your answer? Check your working.	**40 minutes**

Try this

1 A film starts at 6:45 p.m. and finishes at 8:05 p.m. How long does it last? [ ] minutes

2 A film is 1 hour 45 minutes long. It ends at 8:40 p.m. What time did it start? [ ]

3 Buses come once an hour at a quarter past the hour. Mirjana arrives at the bus stop at 8:25 a.m. How long should she have to wait for the next bus? [ ] minutes

Top tip

- Be careful when you write a length of time. 110 minutes is 1 hour and 50 minutes; there are two different unit names. Writing 1.50 hours really means $1\frac{1}{2}$ hours because there is a decimal point used and only one unit name.

Perimeter

To achieve the expected standard, you need to:

★ calculate the **perimeter** of **shapes**, when all side lengths are known or can be easily found.

What you need to know

- The **perimeter** is the total distance around the outside of a shape.
- A **compound shape** is one made from two or more simple shapes (e.g. an L shape is made from two **rectangles**).

Let's practise

What is the perimeter of this shape?

[   ] cm

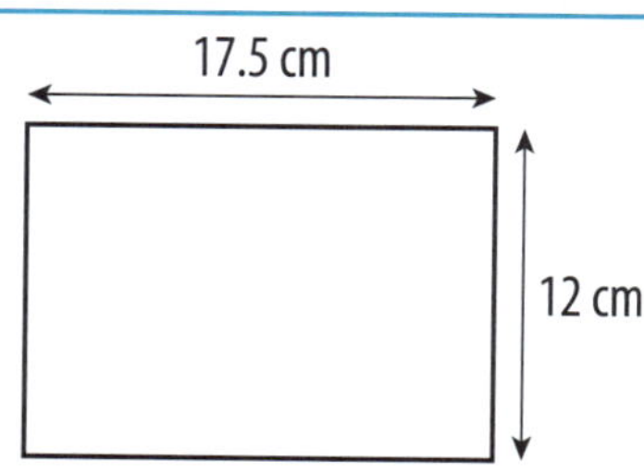

1 Read the question and read it again. What is it asking?

Work out the distance around the shape (the perimeter).

2 Write the length of each side of the rectangle.

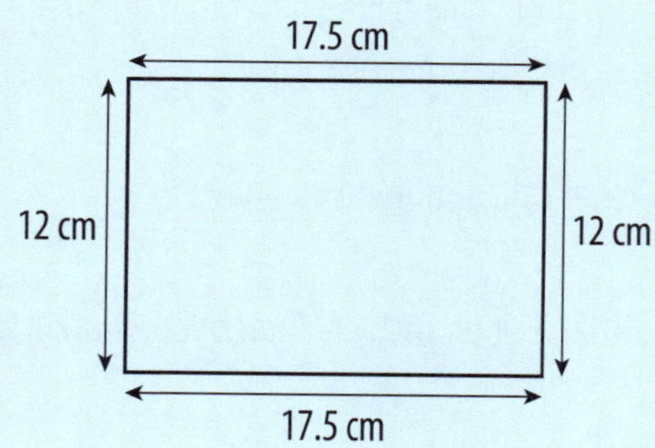

3 Add up all the lengths. Don't forget the units.

12 cm + 12 cm + 17.5 cm + 17.5 cm

4 What is your answer? Double check your working.

The perimeter is 59 cm.

Try this

1 Find the perimeter of this rectangle.

[   ] cm

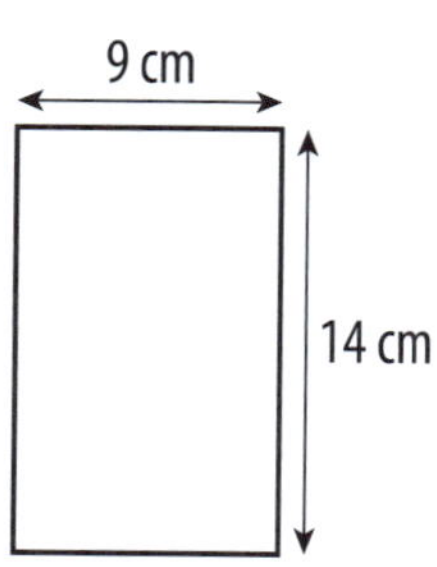

2 A rectangle has a perimeter of 54 cm. Its length is 16 cm. What is the width of the rectangle?

[   ] cm

Estimating the area of irregular shapes

To achieve the expected standard, you need to:

★ estimate the **area** of irregular shapes by counting squares.

What you need to know

- The **area** of a shape is the amount of surface it covers.

Let's practise

This is a plan of a garden pond.
Each square represents $1\,m^2$.

Estimate the area of the garden pond.

☐ m^2

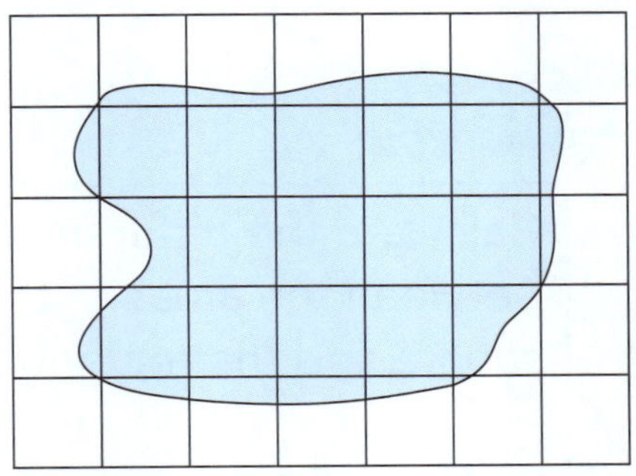

1. Read the question and read it again. What is it asking?
   Estimate the area of the pond and give the answer in square metres.
2. Think of a method you can use.
   I can count the whole squares. There are 12. Next, I need to count any squares that are covered by half or more of the shape as a whole square. There are 3 of these.
3. Count up all the squares and half squares.
   $12 + 3 = 15$
4. What is your answer?
   I estimate the area of the pond to be $15\,m^2$.

Try this

1 Estimate the area of this shape by counting squares. ☐ cm^2

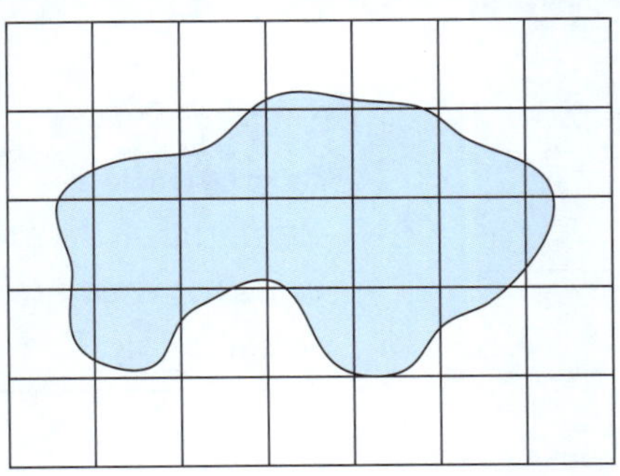

Top tip

- Estimate the area of an irregular shape on a grid by counting whole squares and including squares that are half or more as one whole square.

Area by formula

To achieve the expected standard, you need to:

★ calculate and compare the area of squares and other rectangles

★ use standard units: square centimetres (cm^2) and square metres (m^2).

What you need to know

- The area of a shape is the amount of surface it covers.
- The formula for the area of a rectangle is: Area of rectangle = length × width or A = l × w
- cm × cm = cm^2

Let's practise

Calculate the difference between the areas of the square and the rectangle.

☐ cm^2

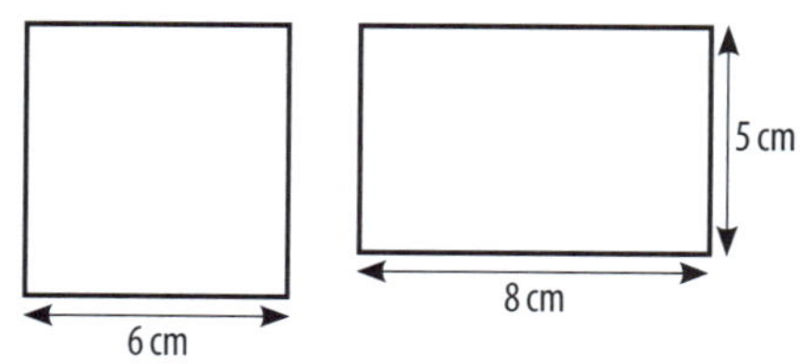

1 Read the question and read it again. What is it asking?	Calculate the area of the square and the rectangle. Then find the difference between them.
2 Start by finding the area of the square. Use the formula A = l × w to calculate the area. Remember to use cm^2 to represent cm × cm.	I know that all sides of a square are equal in length, so each side is 6 cm. A = 6 cm × 6 cm, which is the same as 6^2 = 36 cm^2
3 Now find the area of the rectangle using the same formula as before.	A = 8 cm × 5 cm = 40 cm^2
4 The question asks for the difference between the size of the two areas, so we can use subtraction to help us.	40 cm^2 – 36 cm^2 = 4 cm^2
5 What is your answer? Check your working.	4 cm^2 36 cm^2 + 4 cm^2 = 40 cm^2

Try this

1 Calculate the area of a square with sides of 11 m. ☐ m^2

2 Calculate the area of this rectangle. ☐ cm^2

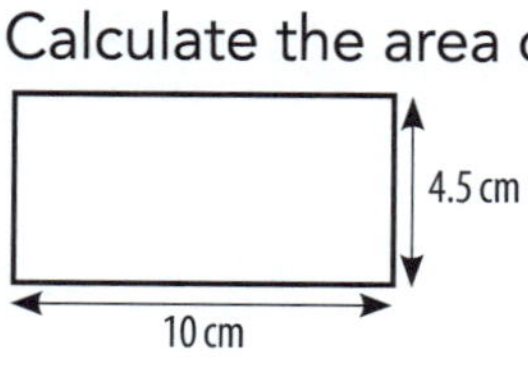

Top tip

- Remember that the units for area are always squares (e.g. cm^2 or m^2).

Drawing lines and angles

To achieve the expected standard, you need to:

★ complete simple 2-D shapes using given lengths and **acute angles** that are multiples of 5 **degrees**.

What you need to know

- An **angle** is a measure of turn and is measured in **degrees** (°).
- A **right angle** is 90 degrees.
- An **acute angle** is less than 90 degrees.

Let's practise

Draw a right-angled triangle with a base of 4 cm and a perpendicular side of 3 cm.

The 4 cm base has been drawn for you.

1. Read the question and read it again. What is it asking?
   Draw a right-angled triangle with side lengths as given.
2. Try to visualise the shape you will be drawing.
   It'll have a right angle in one corner. The longest side will be opposite the right angle.
3. How much space do you have on the grid? Where will you start?
   I'll draw the perpendicular side of 3 cm from the bottom-right corner.
4. Draw the triangle. Check it is accurate.
   There is a right angle between the 3 cm and 4 cm sides.

Try this

1 Copy the diagram to scale and complete the triangle.

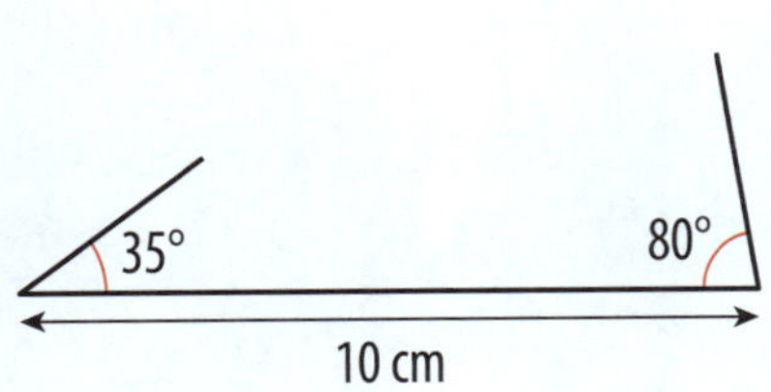

Top tips

- Make sure you use a sharp pencil and a ruler to draw lines.
- Make sure you count the number of degrees from 0 on a protractor.

2-D shapes

To achieve the expected standard, you need to:

- ★ recognise properties of 2-D shapes in terms of number and length of sides, **parallel** sides, type and size of angles
- ★ identify **irregular** and **regular** polygons
- ★ name 2-D shapes according to their properties.

What you need to know

- **Parallel** lines can be straight or curved but they never meet; think of railway tracks.
- **Perpendicular** lines meet at right angles.
- **Polygons** are closed 2-D shapes with straight lines. 

- A polygon is **regular** if all the sides and angles are equal. Otherwise, it is called an **irregular polygon**.

Let's practise

Add the name of each shape in the correct place in the sorting diagram.

	No pairs of parallel sides	1 pair of parallel sides	2 pairs of parallel sides
No perpendicular sides			
At least 1 pair of perpendicular sides			

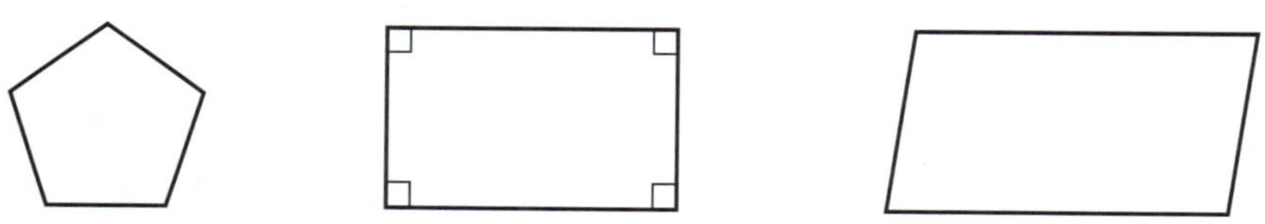

1 Read the question and read it again. What is it asking?

Insert the names of the shapes according to the number of parallel and perpendicular sides they have.

2 Look at one shape at a time.

The pentagon has no parallel sides and no perpendicular sides. Write the name in the correct place in the diagram.

3 What is your answer? Double check it.

pentagon		**parallelogram**
		rectangle

Try this

1 These shapes are drawn on centimetre-squared paper.

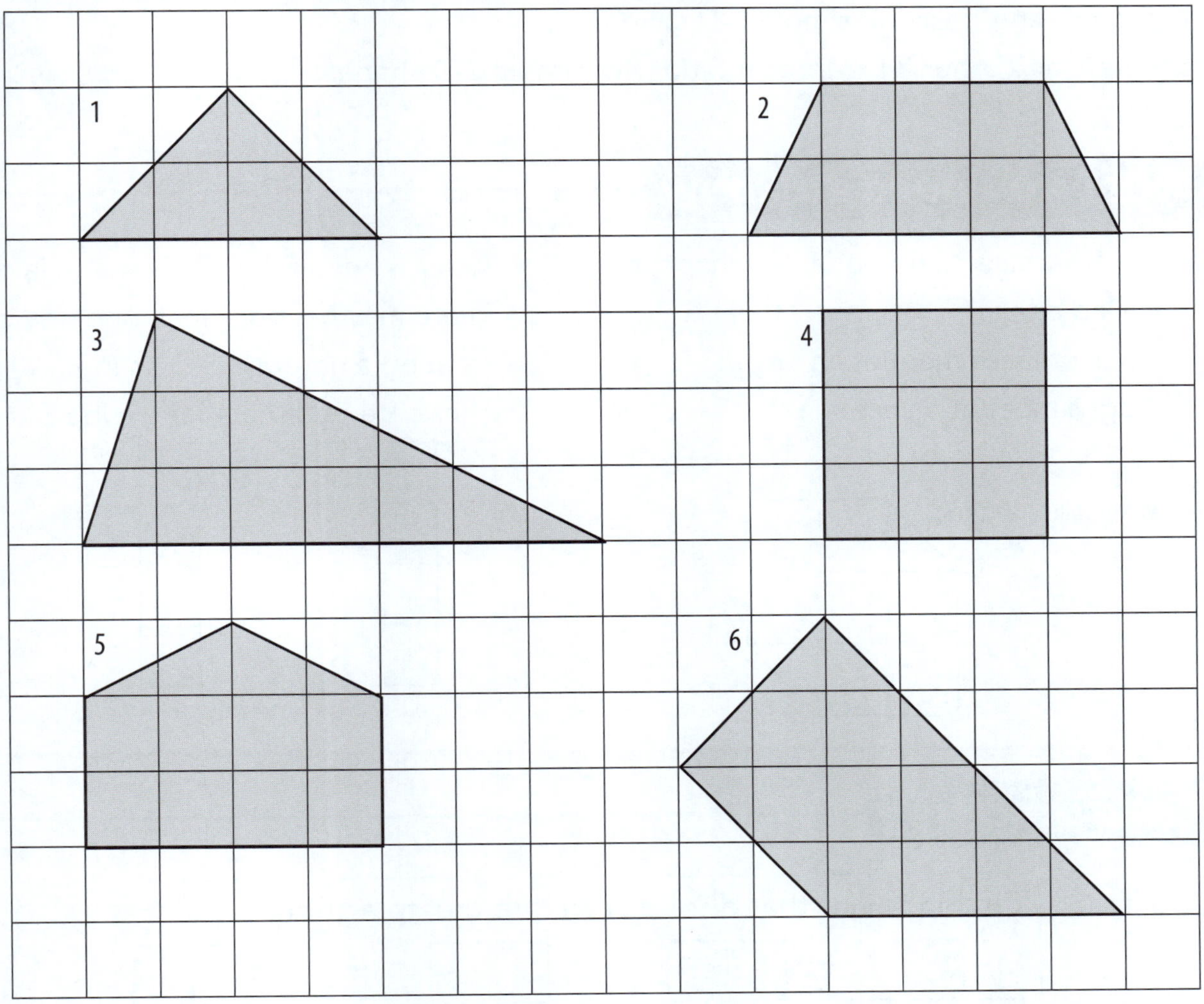

Write the number of each shape in the correct place in the sorting table below.

	No right angles	1 right angle	2 right angles	3 right angles	4 right angles
No pairs of equal sides					
1 pair of equal sides					
2 pairs of equal sides					

Top tips

- Use a ruler to check whether sides are equal.
- Use the gridlines to check whether sides are perpendicular or parallel.

3-D shapes

To achieve the expected standard, you need to:

- ★ recognise properties of 3-D shapes – **faces**, **edges** and **vertices**
- ★ recognise and describe simple 3-D shapes
- ★ use **nets** and other 2-D representations of simple 3-D shapes.

What you need to know

- Each surface or side of a 3-D shape is called a **face**.
  Pairs of faces meet at an **edge**.
  Edges meet at a **vertex**.

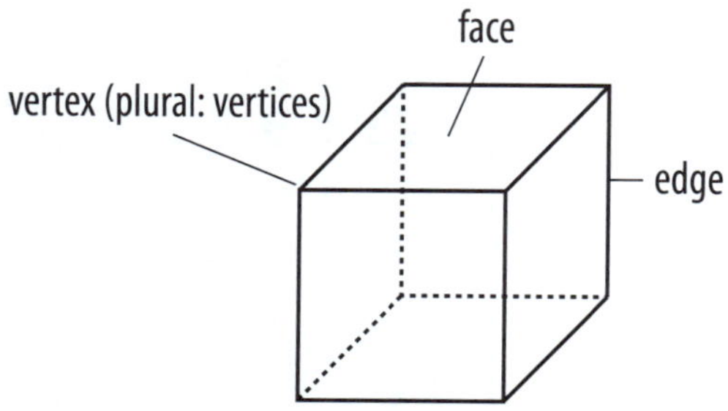

- A **net** is an unfolded 3-D shape laid flat.
  There may be more than one way of creating a net for a 3-D shape.
  These are both nets for a cube.

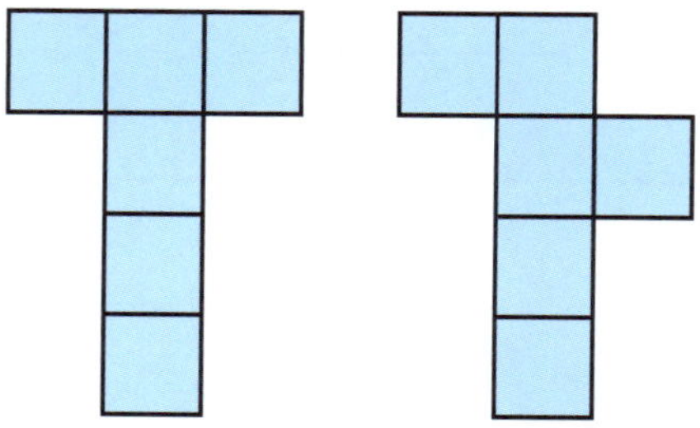

Let's practise

Circle the net that does **not** fold to create a cube.

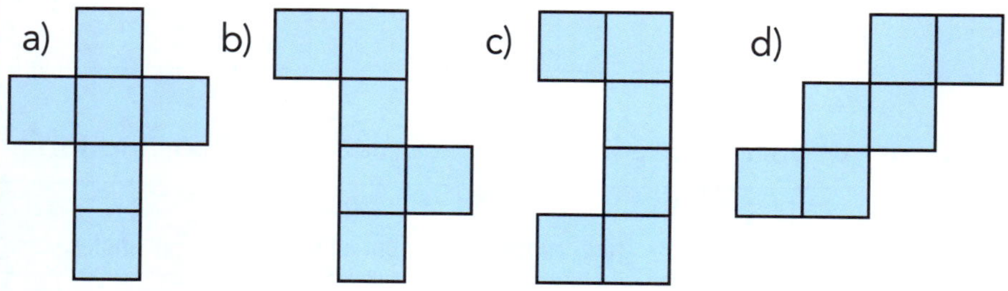

1 Read the question and read it again. What is it asking?	**One of the four nets does not make a cube. I have to decide which one.**
2 Visualise a cube. Consider each option. If you pin one square down, could you fold up the other squares to create a cube?	**This works for all but net c). Here, the four middle squares wrap around, but the other two are on the same side.**
3 What is your answer? Double check it.	**I will circle net c).**

Try this

1 Circle the net that does **not** fold to create a cube.

a)

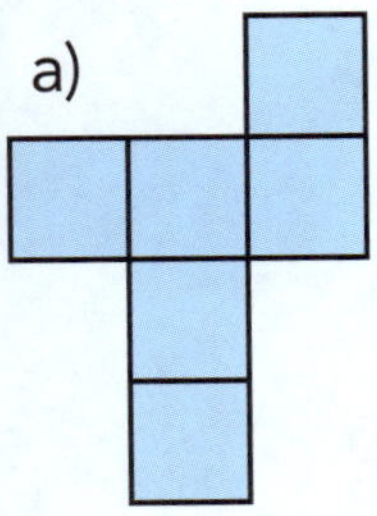

b)

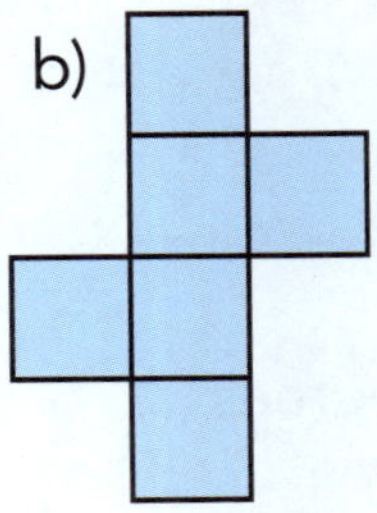

c)

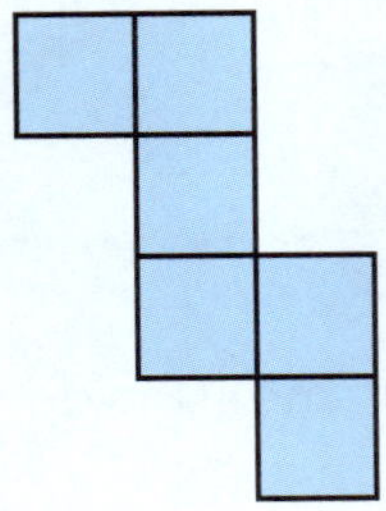

d)

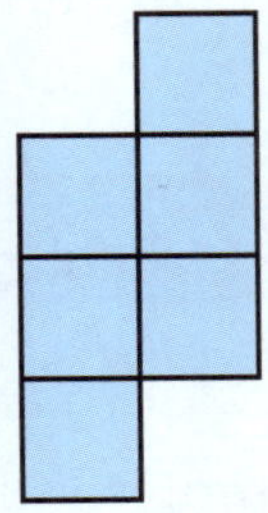

2 Identify these 3-D shapes from their 2-D representations.

a)

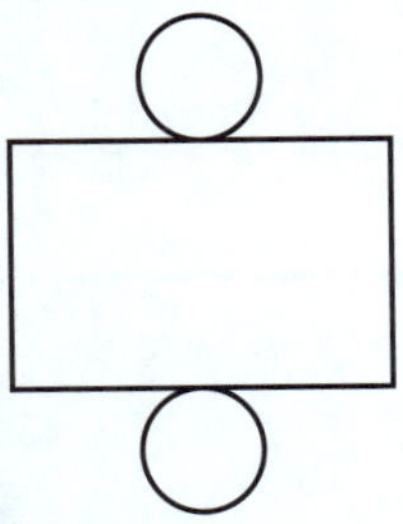

b)

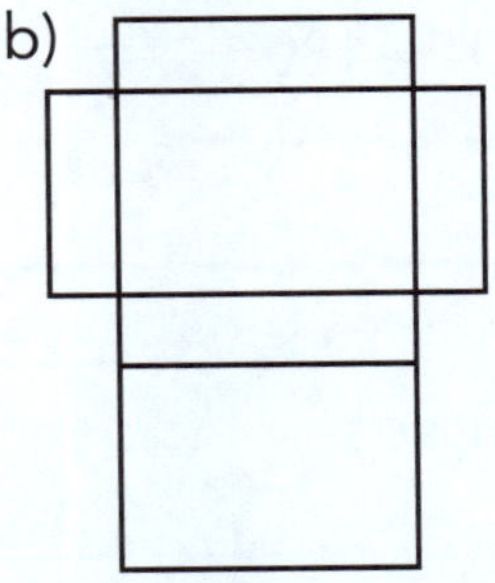

c)

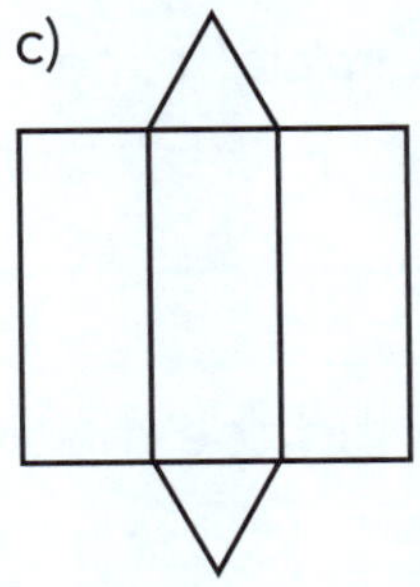

3 Tick any nets of a cuboid.

A

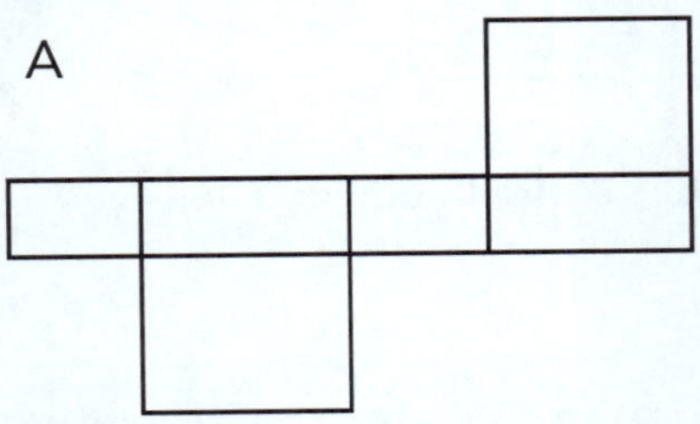

B

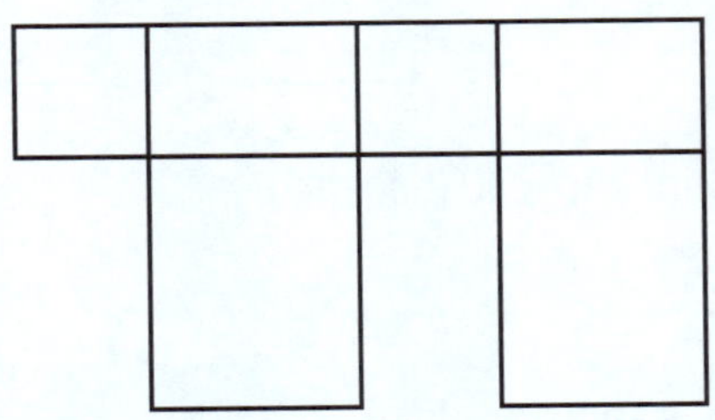

C

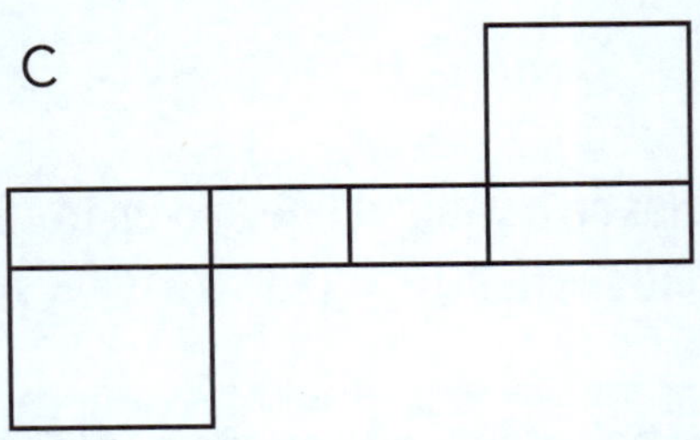

D

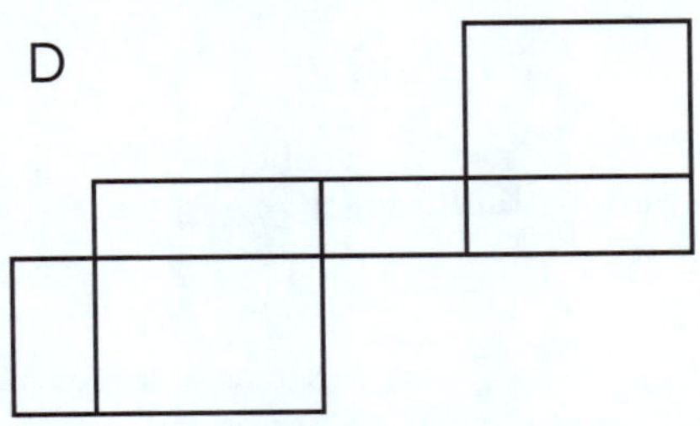

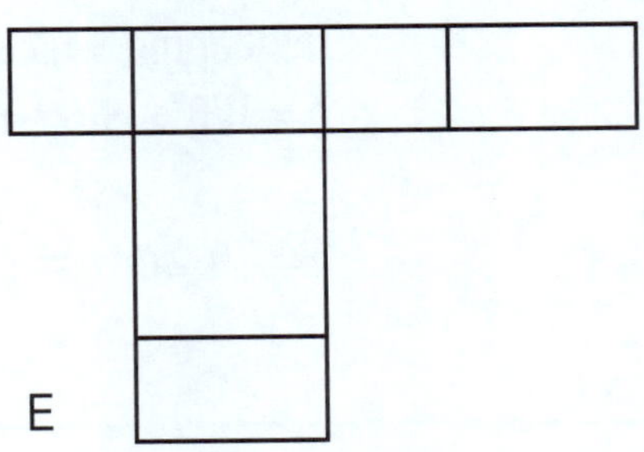

E

Top tip

- Ensure you haven't 'forgotten' a face that may be hidden from view when counting.

Angles and degrees

To achieve the expected standard, you need to:

★ recognise angles where they meet at a point or on a straight line
★ find missing angles at a point and along a straight line.

What you need to know

- An angle is a measure of turn and is measured in degrees (°).
- There are 360° in a complete turn.
- Angles at a point add to 360°.
- A right angle is a quarter turn. $\frac{1}{4}$ of 360° = 90°.
- Two right angles make a straight line. 2 × 90° = 180°.

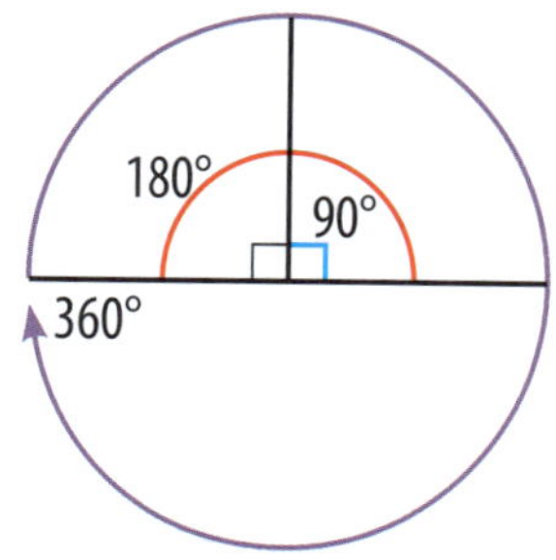

Let's practise

Find the missing angles.

a = ☐ °

b = ☐ °

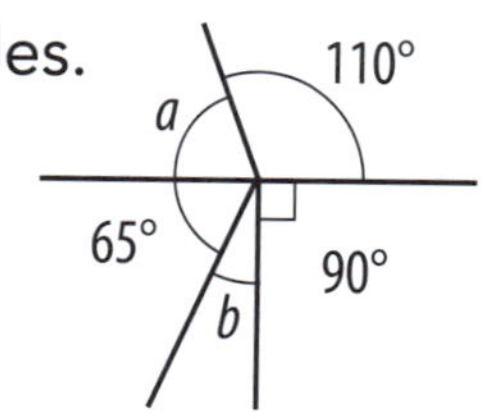

1	Read the question and read it again. What is it asking?	Work out the size of angles *a* and *b*.
2	What can you say about angle *a*?	Angle *a* is on a straight line with another angle of 110°.
3	What do you know about straight lines?	The angles on a straight line add up to 180°, so $a + 110° = 180°$. $a = 180° - 110° = 70°$.
4	What can you say about angle *b*?	*b* is on a straight line with an angle of 65° and a right angle, which is 90°. So $b + 90° + 65° = 180°$. $b = 180° - 90° - 65° = 25°$.
5	What is your answer? Check your working.	$a = 70°$ and $b = 25°$

Try this

1 Calculate angle *a*.

Do not use a protractor.

a = ☐ °

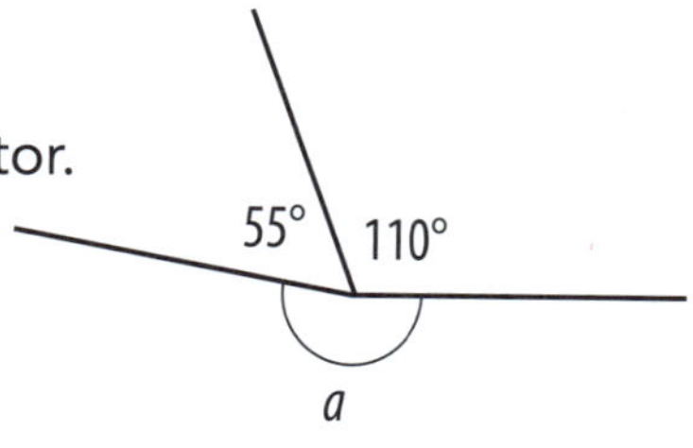

Top tips

- Look for straight lines in a problem as the angles on them add up to 180°.
- Don't forget to put the units in your answer.

Angles in triangles

To achieve the expected standard, you need to:

★ find unknown angles in different types of triangles.

What you need to know

- The angles in a triangle always add to 180°.
- An **equilateral triangle** has three equal angles.
- An **isosceles triangle** has two equal angles.
- A **right-angled triangle** has a right angle at one corner.

Let's practise

What is the size of the other angles in this isosceles triangle?

☐ °

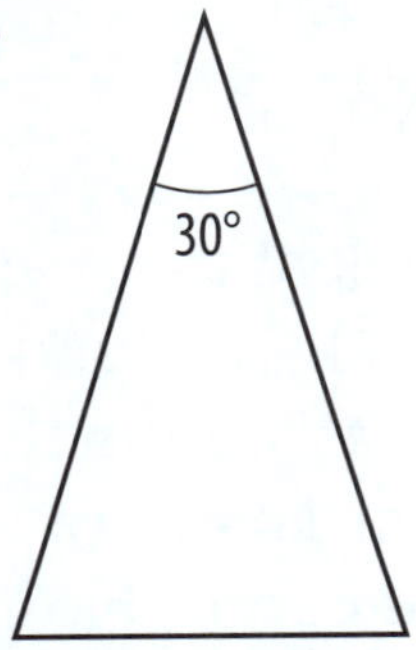

1. Read the question and read it again. What is it asking?
   Work out the size of the unmarked angles.
2. What do you know about the sum of angles in a triangle?
   Angles in a triangle add up to 180°.
3. What do you know about isosceles triangles?
   Isosceles triangles have two sides the same length and two equal angles too.
4. To work out the size of the unmarked angles, first subtract the one you know from 180°. Then divide this number into two equal angles.
   180° – 30° = 150°
   150° ÷ 2 = 75°
5. What is your answer? Check your working and units.
   My answer is 75°.

Try this

1 In a right-angled triangle, one angle is 50°.

What are the other two angles?

☐ ° and ☐ °

Top tip

- Mark any angles you already know on diagrams, even if they are not the answer because they can help you calculate the answer.

Coordinates

To achieve the expected standard, you need to:

- ★ describe the positions on a 2-D coordinate grid using axes with equal scales in the first quadrant and become more confident in plotting points in all four quadrants
- ★ use **coordinates** to complete a given rectangle.

What you need to know

- The **coordinates** of a point (*x*, *y*) describe where it is on a coordinate grid.
- The ***x* coordinate** gives the position along the ***x* axis** (horizontal line).
- The ***y* coordinate** gives the position along the ***y* axis** (vertical line).

Let's practise

Look at this map of a theme park. What are the coordinates of the pirate ship?

(,)

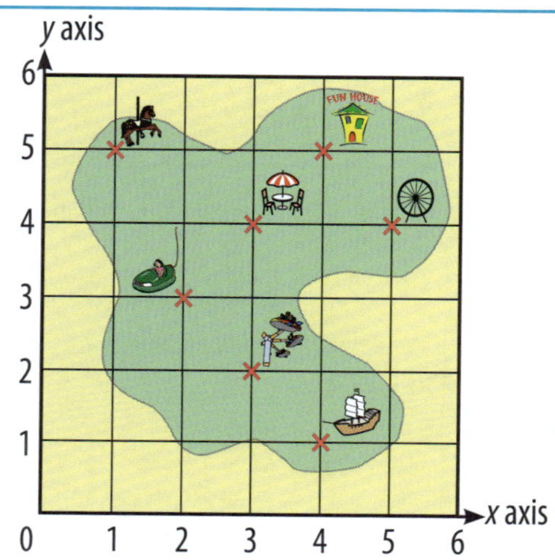

1. Read the question and read it again. What is it asking?

   Describe the position of the pirate ship as *x* and *y* coordinates.

2. Find the pirate ship.

   It's in the bottom right-hand of the park.

3. You need to give its coordinates. From zero, find how far along the *x* axis the ship is positioned. Then find out how far along the *y* axis it is.

   It's 4 across, and 1 up.

4. What is your answer? Write it as a coordinate pair, putting *x* first and *y* second. Double check your answer.

   My answer is (4,1).

Try this

1 Find the coordinate of the café. (,)

2 Find the coordinate of the horse merry-go-round. (,)

Let's practise

Mark a point D to make a rectangle ABCD. What are the coordinates of point D?

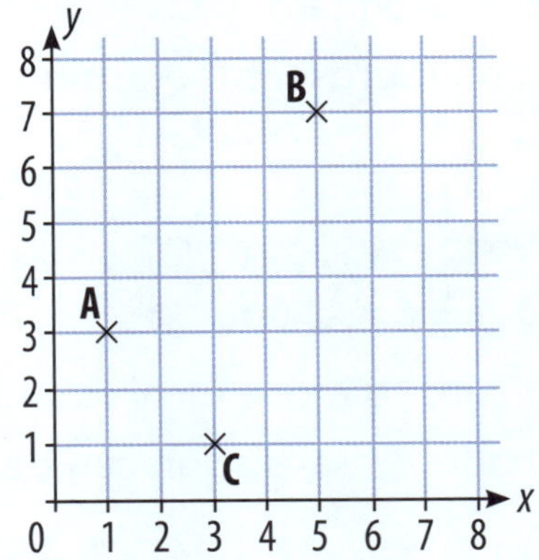

1 Read the question and read it again. What is it asking?

Make a rectangle and state the coordinates of point D.

2 Study the points given. Look for clues for the shape you need to draw.

There are three points. Joining A to B and A to C makes a right angle.

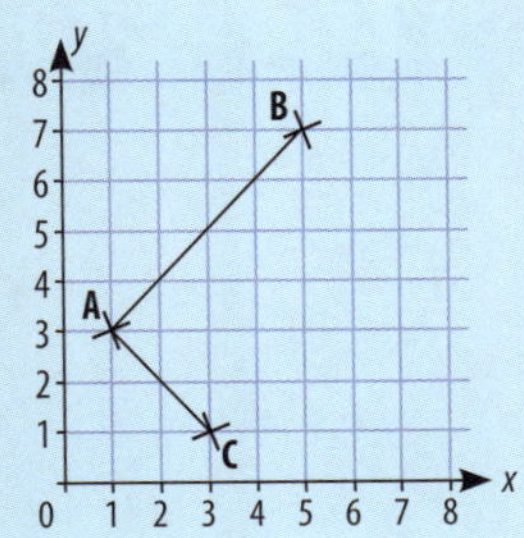

3 How will you complete the shape?

I need right angles at B and at C, with lines parallel to AC and AB.

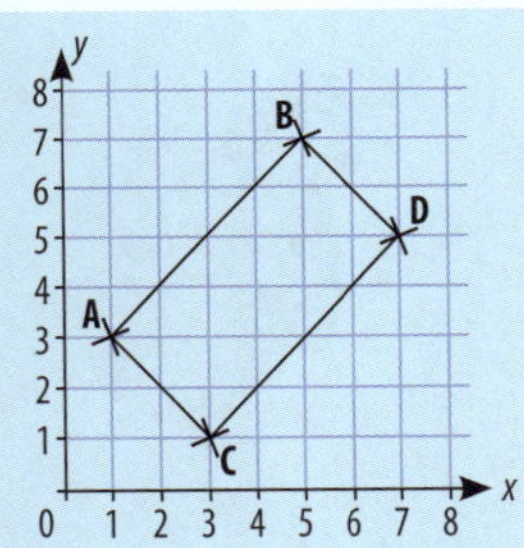

4 Check where the new point is.

The lines meet at (7,5) so that is point D.

Try this

3 The coordinates (1,4) (4,7) (5,0) are three vertices of a rectangle.

What are the coordinates of the fourth vertex? (,)

Top tips

- Always read across the *x* axis and then up or down the *y* axis.
- Write the *x* coordinate before the *y* coordinate like this: (*x*, *y*).

Translations

To achieve the expected standard, you need to:

★ identify, describe and represent the position of a shape following a **translation**.

What you need to know

- A **translation** means to move or slide without turning and without changing the shape. A translation can be:

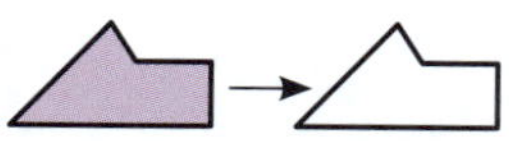

horizontal (left/right)

vertical (up/down)

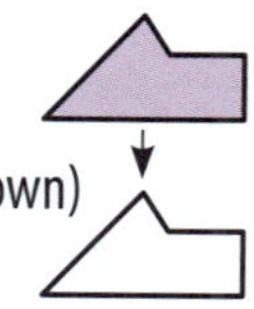

horizontal and vertical

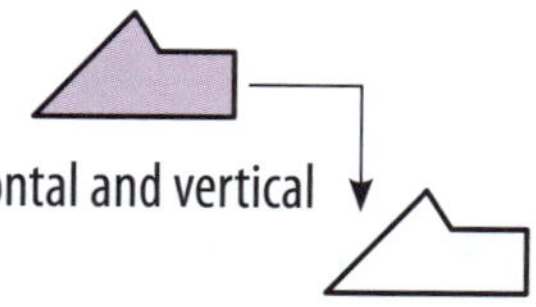

Let's practise

Sketch the position of the shape after a translation of 3 squares right and 4 squares down.

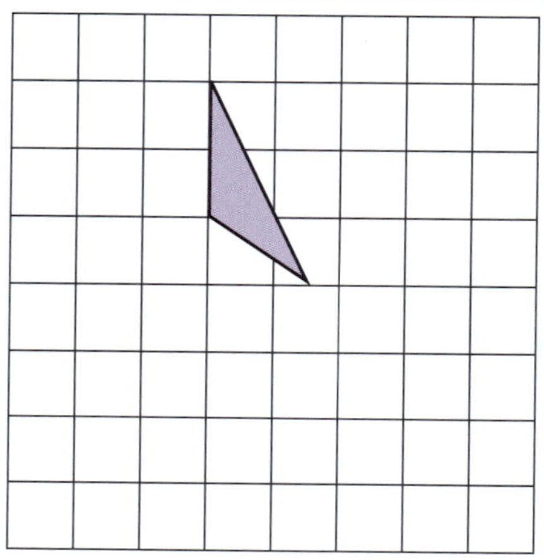

1 Read the question and read it again. What is it asking?

I have to draw another triangle on the grid. It has to be the same shape and size as the original one, but moved 3 squares to the right and 4 squares down.

2 Look at the top vertex of the triangle. Where will it be when the triangle is translated?

3 squares right and 4 squares down.

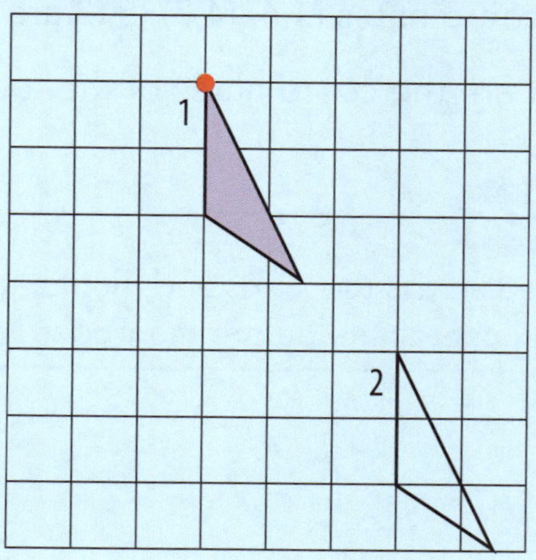

3 Check that the sides are the same length as in the original triangle.

Try this

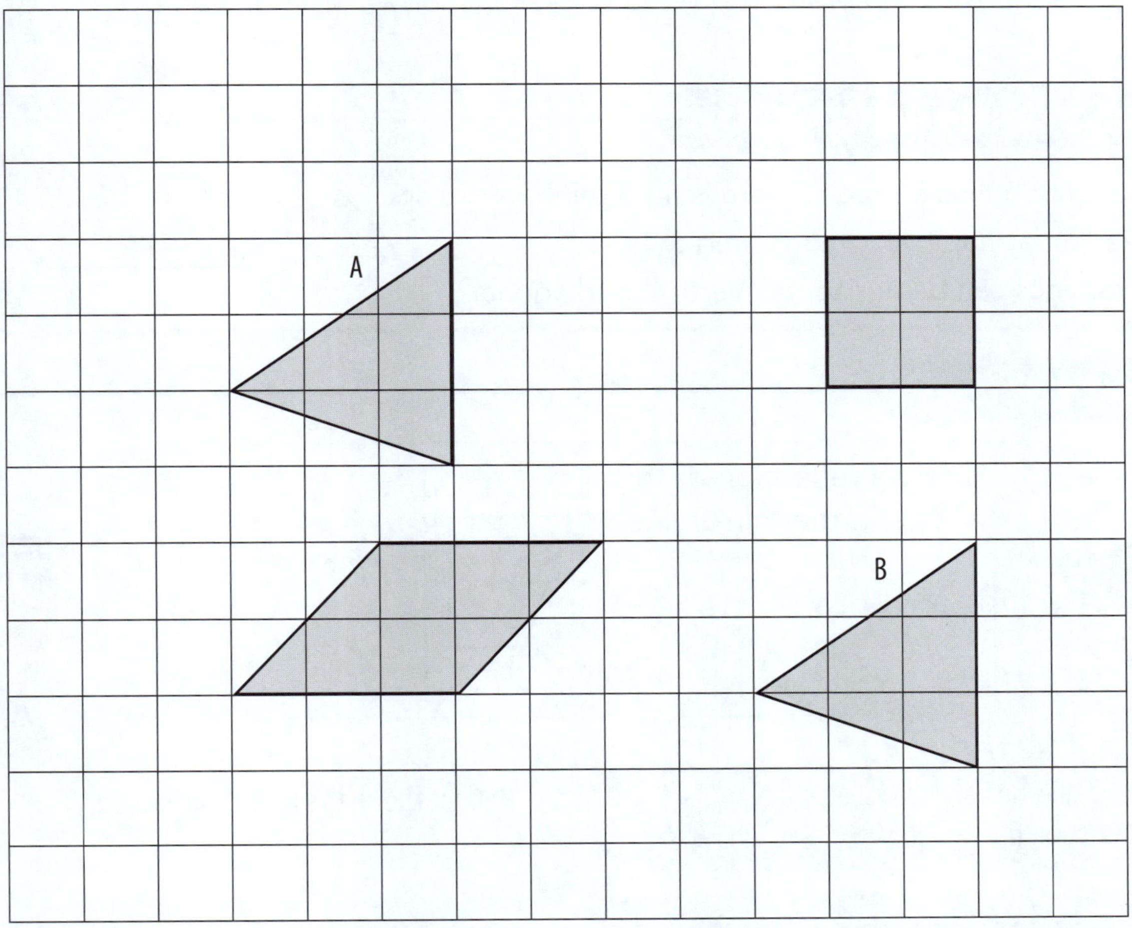

1 The square is translated 4 units left and 6 units down.
Sketch its new position.

2 The parallelogram is translated 1 unit right and 1 unit up.
Sketch its new position.

3 Triangle A has been translated to Triangle B.
The triangle has moved ☐ squares to the right and ☐ squares down.

Top tip

- Check that the shape has not changed in orientation or in size, and that all the sides are still the same **length**.

Reflections

To achieve the expected standard, you need to:

★ identify, describe and represent the position of a shape following a **reflection**.

What you need to know

- A **reflection** flips a shape. There is no turning or sliding.
- To reflect a shape you need a **mirror line**.
- Mirror lines can be horizontal, vertical or diagonal.

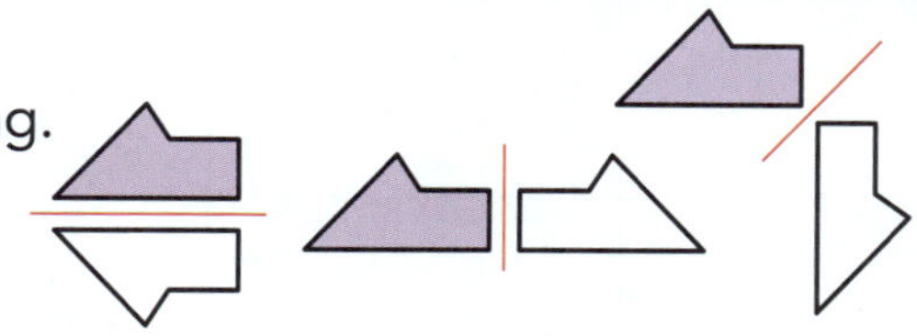

Let's practise

Draw a reflection of this shape in the mirror line.

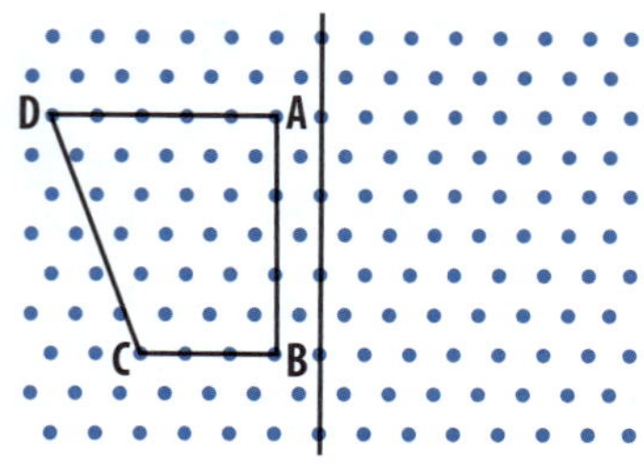

1 Read the question and read it again. What is it asking?

Reflect the shape in the mirror line.

2 Notice where each corner is positioned. In the reflected shape, each corner will be the same distance from the mirror but on the other side of the mirror line.

Corner A is one dot away from the mirror line. Corner D is on the same line.

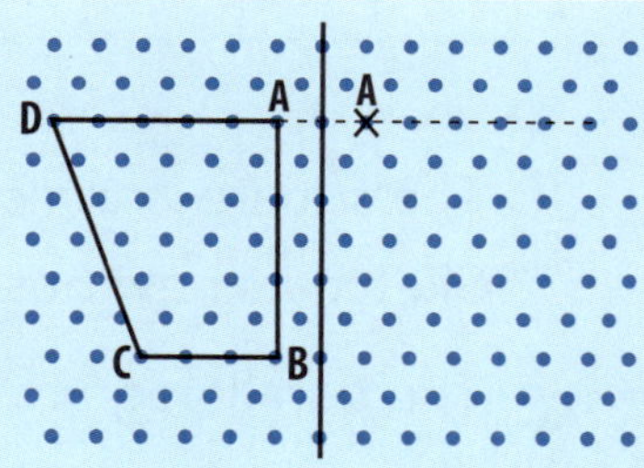

3 Repeat this for the other points. Then use a ruler to join the corners. Check that the sides are all the same length as in the original shape.

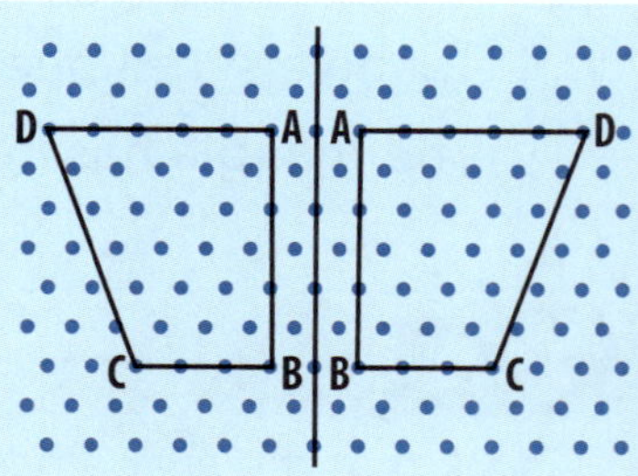

Try this

Reflect these shapes in the mirror lines.

1

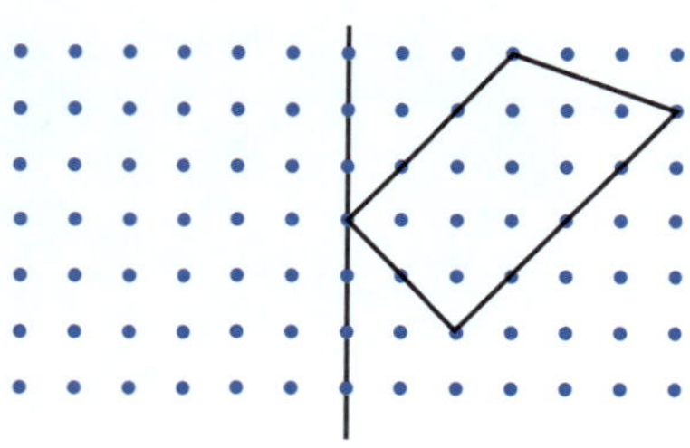

2

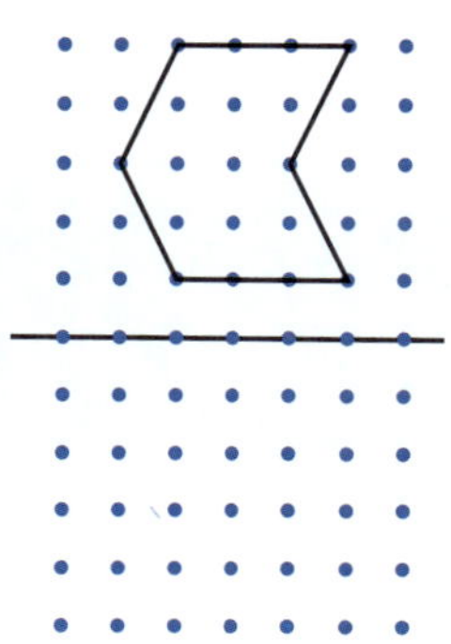

Top tip

- Try using a mirror to check the reflection is accurate.

Reflective symmetry

To achieve the expected standard, you need to:

★ compare and classify 2-D shapes in terms of **reflective symmetry**.

What you need to know

- A shape has **line symmetry** (sometimes called **reflective symmetry**) if you can fold it in half along a line so that the two halves match exactly. The folding line is called the **line of symmetry**.

Let's practise

Tick the shapes that have reflective symmetry.

a) 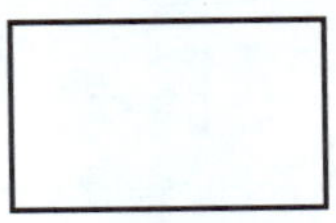

b) 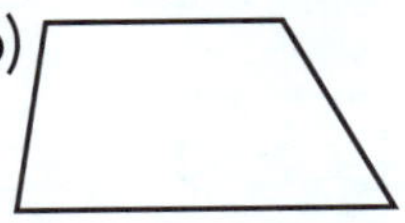

c) 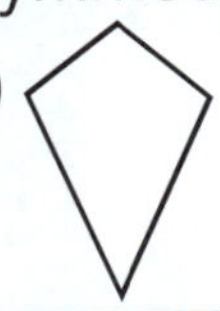

1 Read the question and read it again. What is it asking?	Identify the shapes that have reflective symmetry.
2 It can help to look at each shape and identify it first.	a) is a rectangle, b) is a **trapezium**, c) is a kite.
3 Which shapes can be cut into two identical halves (e.g. by a diagonal or a perpendicular line)?	The diagonals of the rectangle do not cut it into two identical halves, but the perpendicular bisectors of the sides do. 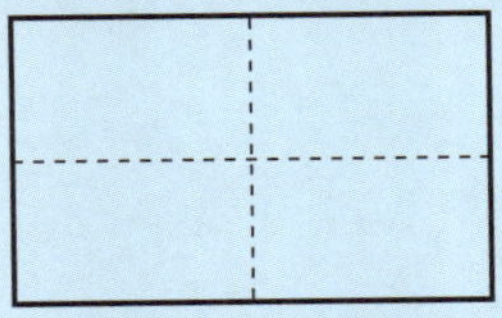 The trapezium has no symmetry. Only the vertical diagonal of the kite cuts it into two halves. 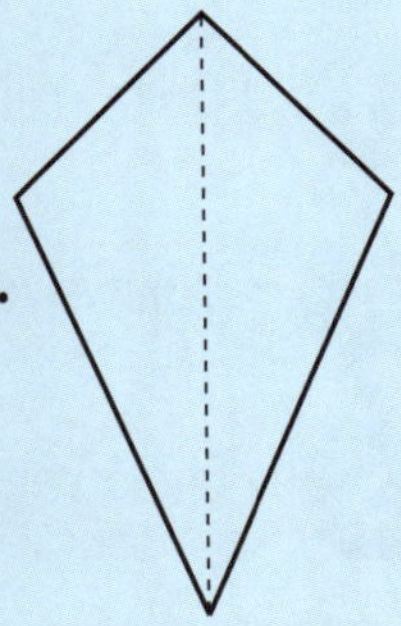
4 What is your answer?	I tick a) and c).

Try this

1 How many lines of symmetry does each shape have?

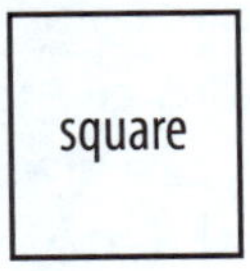

☐

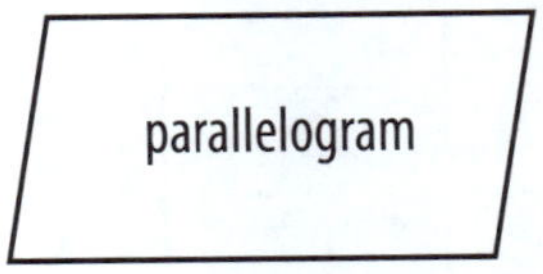

☐

☐

Top tip

- Try turning the shapes around to see them in a different position.

Tables

To achieve the expected standard, you need to:

★ complete, read and interpret information in **tables**.

What you need to know

- **Tables** provide one way of presenting data, in rows and columns.

Let's practise

The table shows the number of pupils at Park Primary School.

	Reception	Key Stage 1	Key Stage 2	**Total**
Boys	11		104	
Girls	13	69		192
Total		131		

How many girls are in Key Stage 2 at Park Primary School? ☐

1. Read the question and read it again. What is it asking?
   Work out the number of girls in Key Stage 2 at the school.
2. How will you do this?
   The total number of girls in the school is given (192). There are 13 girls in Reception and 69 girls in Key Stage 1. I need to add 13 and 69 and subtract this total from 192.
3. Do the calculation.
   $192 - (13 + 69) = 192 - 82 = 110$
4. What is your answer? Don't forget the units.
   My answer is 110 girls.
5. Check your calculation by adding.
   $13 + 69 + 110 = 192$

Try this

1. Use the Park Primary School data. How many children are in the school altogether? ☐
2. The table below shows the number of flights from Newcastle airport to UK cities during two weeks. Fill in all the missing numbers in this table.

	London	Bristol	Belfast	Manchester	**Total**
Week 1	32	15		8	
Week 2			16		76
Total	68	27			147

Top tip

- Remember columns are always vertical (up/down) and rows are always horizontal (left/right).

Pictograms

To achieve the expected standard, you need to:

★ interpret and present data using **pictograms**.

What you need to know

- **Pictograms** use images to represent numbers.

Let's practise

This pictogram shows what fillings children like in their sandwiches.

Which is the least popular filling? ______

How many children like this filling? [ ]

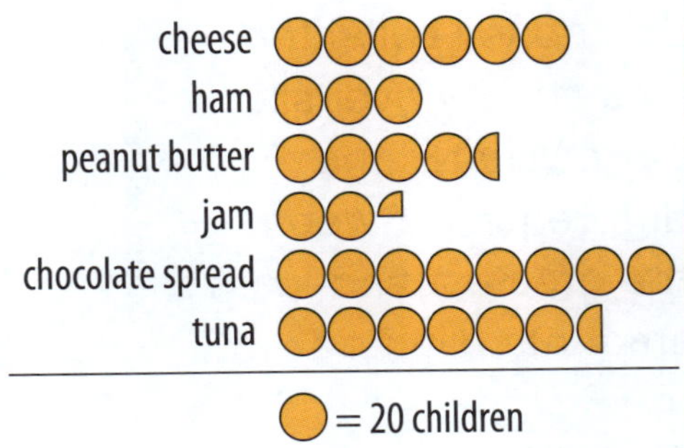

1	Read the question and read it again. What is it asking?	From the data, find the least popular filling and work out how many children like this filling.
2	The most popular filling has the greatest number of symbols. The least popular has the fewest number of symbols.	Jam has only $2\frac{1}{4}$ symbols so must be the least popular filling.
3	Check what each symbol represents.	Each symbol represents 20 children.
4	Work out the number of children who like jam sandwiches.	There are $2\frac{1}{4}$ symbols in the jam column. Each symbol stands for 20 children. $20 \times 2\frac{1}{4} = (20 \times 2) + (20 \times \frac{1}{4}) = 40 + 5 = 45$
5	What is your answer? Check your working.	Jam is the least popular filling. 45 children like jam sandwiches.

Try this

Now answer these questions for the same pictogram.

1 What is the second most popular filling? ______

2 How many more children like chocolate spread than peanut butter? [ ] children

3 How many children in total like cheese or ham? [ ] children

Top tip

- Always check what the value is for each symbol.

Bar charts

To achieve the expected standard, you need to:

★ complete, read and interpret information presented in **bar charts**.

What you need to know

- In a **bar chart** there is one bar per category of data.

Let's practise

This bar chart shows the money collected by a charity over six days. What is the difference between the greatest and least amount collected?

£ [ ]

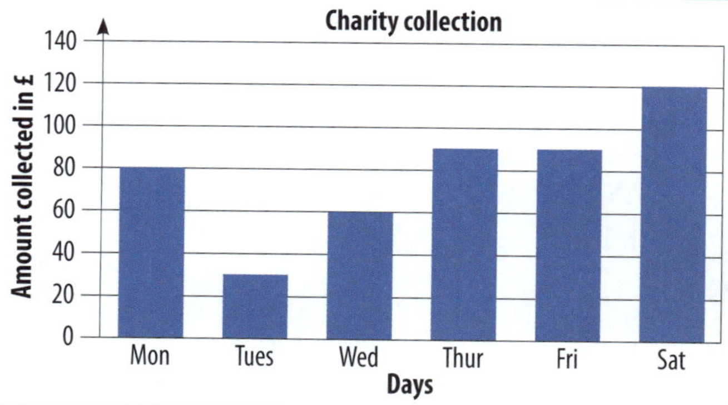

1	Read the question and read it again. What is it asking?	**Identify the greatest and least amounts and subtract to find the difference.**
2	The greatest amount is the tallest column (Sat).	**The bar is on 120.**
3	The least amount is the shortest column (Tues). It does not match a horizontal line. You will need to estimate the amount.	**The bar comes midway between 20 and 40, so 30 would be a reasonable estimate.**
4	Complete the subtraction to find the difference.	**120 – 30 = 90**
5	What is your answer? Check you have included the units.	**£90**

Try this

Top tip

- Always read any scales on the axes carefully.

Answer these questions using the same bar chart.

1 On which **two** days was the same amount of money collected? ____________________

2 What is the difference between the amount of money collected on Monday and Friday?

[ ]

3 How much money was collected on Friday and Saturday altogether? [ ]

Pie charts

To achieve the expected standard, you need to:

★ interpret simple **pie charts**.

What you need to know

- A **pie chart** is one way of presenting data.
- A pie chart shows the proportion of each category, through a share of the 'pie'.
- Proportion can be given as a fraction or as a percentage.

Let's practise

This pie chart shows the ages of the population living in a town.

a) What fraction of the population is aged over 75? ☐

b) What fraction of the population is aged 21 and under? ☐

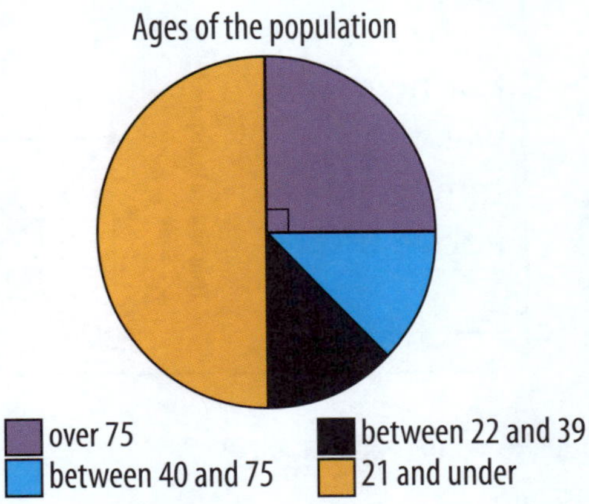

1	Read the question and read it again. What is it asking?	**Read the pie chart and answer two questions.**
2	Read the key carefully. Match the sections on the pie chart to the colours on the key.	**For part a) I need people over 75 so that is the purple section.**
3	The question asks for the fraction of the population.	**A quarter of the circle is shaded purple.**
4	Now look at part b).	**Age 21 and under is the orange part of the pie chart. That's half of it.**
5	What is your answer?	**a) $\frac{1}{4}$ b) $\frac{1}{2}$**

Try this

Answer these questions using the same data.

1 What fraction of the population is aged between 22 and 39? ☐

2 What fraction of the population is aged 75 or under? ☐

3 What percentage of the population is aged over 75? ☐ %

Top tips

- Be careful to match the key (what each colour means) to the pie chart.
- Check to see if you are asked for the fraction or the percentage.

Line graphs

To achieve the expected standard, you need to:
★ interpret **line graphs** to solve problems.

What you need to know

- In a **line graph**, readings are taken and recorded as separate points.
- The separate points can be joined to show change.

Let's practise

This graph shows the temperature over one day.

For how long was the temperature above 0°C? ☐

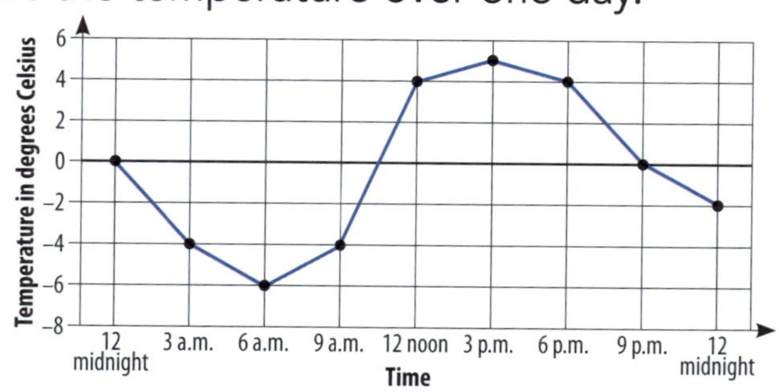

1 Read the question and read it again. What is it asking?

Work out for how long the temperature was above 0°C.

2 Study the graph.

The temperature first rises above the 0°C halfway between 9:00 a.m. and 12 noon (10:30 a.m.). The temperature falls below 0°C at 9:00 p.m.

3 Calculate the difference between 10:30 a.m. and 9:00 p.m. Do this by counting on.

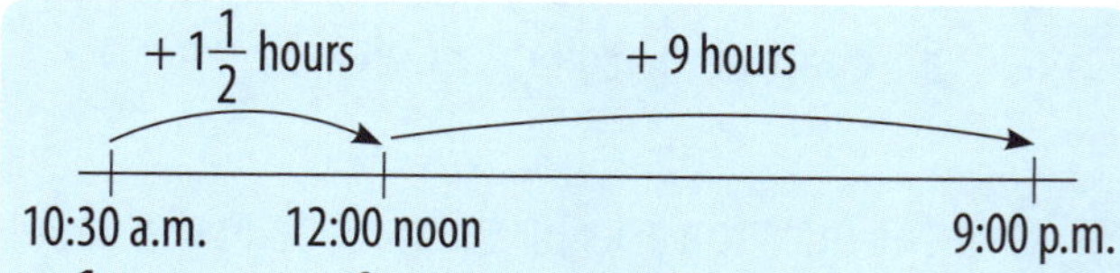

$1\frac{1}{2} + 9 = 10\frac{1}{2}$

4 What is your answer? Check you have included the units.

$10\frac{1}{2}$ hours

Top tips

- When reading graphs, follow the lines carefully. Using a ruler can help.
- It's a good idea to make an estimate by thinking of dividing the gap into a fraction (e.g. the point is about $\frac{1}{2}$ of the way between 4 and 6, which makes the value about 5).

Try this

Use the temperature graph to answer these questions.

1 At what time was the temperature 5°C? ☐

2 Work out the difference in the temperature between 6:00 a.m. and 6:00 p.m. ☐ °C

Mean

To achieve the expected standard, you need to:

- ★ calculate and interpret the **mean** as an **average** for simple sets of discrete data.

What you need to know

- An **average** is a central value of a set of data.
- One example of an average is the **mean**. To find the mean, add all the amounts and divide the answer by the number of amounts.
- **Discrete data** gives information about the number of things we can count and only refers to whole number values, e.g. if the graph was about children, the only numbers could be whole numbers, because you couldn't have $3\frac{1}{2}$ children.

Let's practise

What is the mean of this set of data?

2 2 4 6 10 12 ☐

1	Read the question and read it again. What is it asking?	**Find the mean of some data items.**
2	Count the number of data items.	**There are 6 data items.**
3	Add all the amounts.	**$2 + 2 + 4 + 6 + 10 + 12 = 36$**
4	Divide the sum of the amounts by the number of data items. Check that your answer looks sensible.	**$36 \div 6 = 6$ 6 is bigger than 2, the smallest data item, and smaller than 12, the largest data item. It is somewhere in between, which looks about right. So the answer is 6.**

Try this

1. Find the mean of this set of data: 12 13.5 4.5 10 5 ☐
2. Max spends a mean of £4.50 on his lunch each day from Monday to Friday. How much does he spend altogether on these five days? ☐
3. The mean of five numbers is 15
   Four of the numbers are 12
   What is the fifth number? ☐

Top tip

- The mean is always larger than the smallest data item and smaller than the largest data item.

Whole number problems

To achieve the expected standard, you need to:

- ★ solve problems with two or three steps involving addition, subtraction, multiplication and division and a combination of these operations
- ★ use your knowledge of the four operations to reason and to solve problems, including puzzles that are not set in a context.

Let's practise

A factory packs tins of soup in boxes. There are 48 tins of soup in a box. The factory has 6,816 boxes of soup in stock and 8,544 tins that need to be packed into boxes.

How many boxes will there be in total? ☐

1	Read the question and read it again. What is it asking?	**I need to find out how many boxes of soup there are altogether.**
2	Divide 8,544 tins of soup by 48. Next, add the number of boxes together.	**$8{,}544 \div 48 = 178$** **$6{,}816 + 178 = 6{,}994$**
3	When you have an answer, check it.	**$6{,}994 - 6{,}816 = 178$** **$178 \times 48 = 8{,}544$**
4	Check your answer using the inverse operation. Write it in the box. Don't forget to include the units.	**6,994 boxes**

Try this

1 Two types of plane fly between London and New York.
One type of plane has 354 seats and has 25 flights each week.
The other type of plane has 423 seats and has 18 flights each week.
How many passengers could travel from London to New York each week?

☐ passengers

2 Harry plays a computer game. He scores 2,500 points for collecting 10 gold rings and 1,500 points for collecting 10 silver rings.
Harry collects 70 gold rings and 50 silver rings.
How many points does he collect altogether?

☐ points

Problems with fractions, decimals and percentages

To achieve the expected standard, you need to:

★ solve problems and reason about fractions, decimals and percentages.

Let's practise

Nishi sits three tests at school.
Her marks are 17 out of 20 for spelling, $\frac{43}{50}$ for English and 87% for maths.
Which is Nishi's best score? ______________

1. Read the question and read it again. What is it asking?

   Find the best or greatest score. This will mean changing the fractions and percentages so they are equivalent.

2. Note key numbers.

   - Spelling 17 out of 20 or $\frac{17}{20}$
   - English $\frac{43}{50}$
   - Maths 87%

3. Turn the fractions into percentages. First turn the denominators of the fractions into hundredths.

   Spelling 17 out of 20
   Multiply numerator and denominator by 5 $= \frac{85}{100}$
   English $\frac{43}{50}$
   Multiply numerator and denominator by 2 $= \frac{86}{100}$

4. Turn the fractions into percentages.

   Spelling = 85%
   English = 86%
   Maths = 87%
   Now, I need to compare the answers.

5. Check that your answer looks sensible. Write it in the box.

   87% is the greatest score.
   Nishi's best score is in maths.

Try this

1 Three bottles can each hold 1 litre. Some milk is poured into the bottles. Bottle 1 has $\frac{2}{5}$ litre, bottle 2 has 0.38 litre and bottle 3 is 43% full.

Which bottle has the most milk? Bottle ______

Top tip

- When you are comparing fractions, decimals and percentages, it's best to change them all into hundredths.

Measurement problems

To achieve the expected standard, you need to:

★ solve problems involving the calculation and conversion of units of measure, using decimal notation up to two decimal places where appropriate

★ explain reasoning for problems involving measures.

Let's practise

A rectangle has an area of 72 cm^2 and a perimeter of 76 cm. What is the length and width of the rectangle?

length = ________ width = ________

1	Read the question and read it again. What is it asking?	**Find the length and width of the rectangle.**
2	Picture the problem. You could draw a sketch.	
3	Note key words. Remember the meaning of 'area' and 'perimeter' and how to find the area and the perimeter of a rectangle.	**Area = length × width** **Perimeter = 2 × (length + width)**
4	Note key numbers.	**72 = length × width** **76 = 2 × (length + width)**
5	Decide which operation or operations you will use.	**I need to think of two numbers that multiply to give 72. The same two numbers when added and then doubled must give 76.**
6	Decide on your plan and work through the steps systematically.	**I can make a list of numbers that multiply to 72. Then I can work through these numbers to find a pair that can be added and doubled to give 76.**
7	When you have an answer, check it.	**$36 \times 2 = 72$** **$2 \times (36 + 2) = 76$**
8	Check that your answer looks sensible. Don't forget to include units.	**length = 36 cm, width = 2 cm**

Try this

1 The area of a rectangle is 36 cm^2.
What could the perimeter of the rectangle be? [ ] cm

2 The perimeter of a rectangle is 40 cm.
What could the area of the rectangle be? [ ] cm^2

Top tip

- Make sure you remember how to find **lengths** and widths from areas and perimeters of **rectangles**.

Shape problems

To achieve the expected standard, you need to:

★ solve problems and reason about shapes and their properties.

What you need to know

- The interior angles of a triangle total 180°. The interior angles of a quadrilateral total 360°.
- Angles that are opposite each other on equal sides of an isosceles triangle are equal in size.
- Angles on a straight line total 180°.

Let's practise

This is a rectangle with its two diagonals shown.
An angle of 25° is marked.

Calculate the size of angle x. [ ]°

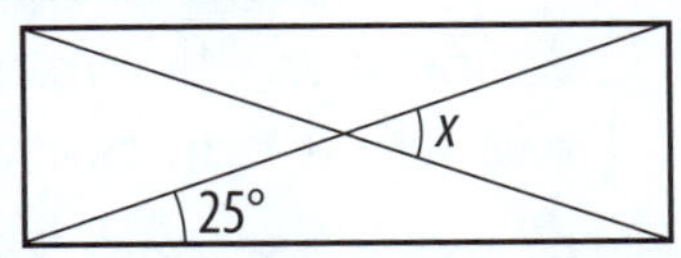

1	Read the question and read it again. What is it asking?	**Calculate the size of the angle marked *x*. Calculate means use arithmetic; don't measure the angle.**
2	Note key words.	**The shape is a rectangle but I've noticed that the diagonals also make four isosceles triangles.**
3	Note key numbers.	**One angle in one of the isosceles triangles is 25°.**
4	Decide which operation or operations you will use.	**To find a missing angle in a triangle or on a straight line, subtract all other known angles from 180°.**
5	Decide on your plan and work through the steps systematically.	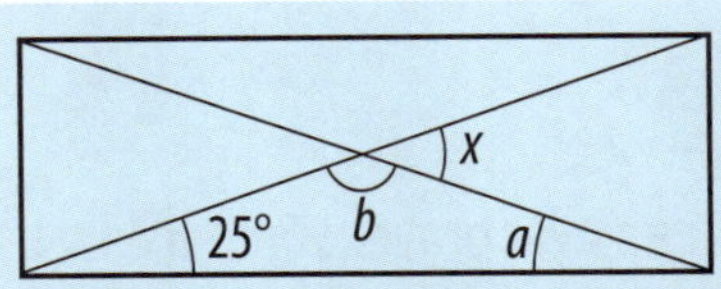 **Angle *a* is also 25°, so angle *b* must be 130° [180 – (25 + 25)]. Angle *x* is 50° [180 – 130].**
6	Check that your answer looks sensible. Write it in the box. Don't forget to include units.	**50°**

Try this

1 Find angles *p* and *q*. *p* = [ ]° *q* = [ ]°

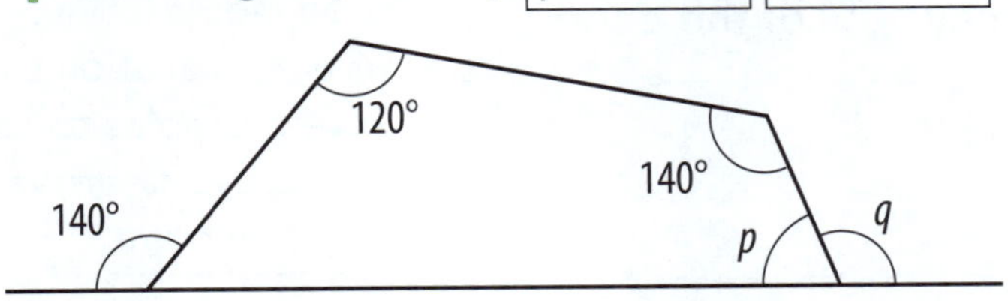

Top tip

- Learn the totals of the angles in degrees in triangles and quadrilaterals, on a straight line and around a point.

Reasoning

To achieve the expected standard, you need to:

★ begin to reason mathematically making simple generalisations, using mathematical language.

What you need to know

- How to predict and test your ideas.
- How to look for patterns and relationships.
- How to give a statement and then prove or disprove it with examples.

Let's practise

Josef says, 'If you add a two-digit number to another two-digit number you can't get a four-digit number.'

Is Josef correct? YES / NO

Give an explanation for your answer. ______________________

____

1 Read the question and read it again. What is it asking?

I need to see if Josef's statement is true or false.

2 Thinking about the statement, if you try to make as large a number as possible, you will need to use the largest two-digit numbers possible.

$99 + 99 = 198$

3 When adding two two-digit numbers, the largest total is 198, which is not a four-digit number.

Making a four-digit number is not possible.

4 What is your answer? Check that your answer proves your idea by using examples.

The largest number that can be made by adding two two-digit numbers is $99 + 99 = 198$ because 99 is the largest two-digit number. Josef is correct and so I circle YES.

Try this

1 Preeya did a survey of 95 people to see how many were left-handed.

Preeya says,'The results show that exactly 10% of the people in the survey are left-handed.'

Explain why Preeya cannot be correct.

____

Top tips

- Try a few different examples to test an idea, not just one.
- Your explanation should give examples to help you prove or disprove the statement.

Glossary

Acute angle An angle that is less than 90°.

Area The surface taken up by a 2-D shape, such as a square or a triangle. Usually, area is measured in squares of the same size, such as centimetre squares (cm^2) or metre squares (m^2).

Bar chart A type of graph that uses blocks or bars to represent different values. The bars can be shown vertically or horizontally.

Capacity The volume of a liquid held in a container; measured in litres or millilitres.

Compound shape A shape made up from other shapes.

Coordinates A pair of numbers that show a position on a grid or a map. The pair of numbers is usually written in brackets and separated by a comma, e.g. (6,3). The first number refers to the number on the horizontal or *x* axis and the second number refers to the number on the vertical or *y* axis.

Degree A degree is a measure for both angles and temperature. The symbol for degrees is °. Measuring an angle is measuring a turn. There are 360° in a full turn, so 1° is $\frac{1}{360}$ of a full turn. For temperature, a degree is the unit of measurement. Two different scales are used: Celsius (°C) and Fahrenheit (°F).

Denominator *See Numerator.* The number on the bottom of a fraction. The denominator shows how many parts are in one whole, e.g. in $\frac{1}{5}$ there are five fifths or parts needed to make 1.

Discrete data Data that can only have certain values, usually only whole numbers, e.g. a graph showing the number of children in a class must only use whole numbers as it is impossible to have half a child.

Edge Where two faces (sides) of a 3-D shape meet.

Equilateral triangle A triangle with three equal sides. The angles in an equilateral triangle are also equal and are always 60°.

Equivalent fraction A fraction that has the same value, e.g. $\frac{1}{5} = \frac{2}{10} = \frac{3}{15} = \frac{4}{20}$. All of these fractions are the same even though they use different numbers.

Face The side of a 3-D shape. Faces are usually flat and are 2-D shapes. When the side of a 3-D shape is curved as in a cone, the side is usually called a 'surface'.

Factor *See Multiple.* A whole number that can be multiplied by another number to give a whole number, e.g. 4 × 5 = 20, so 4 and 5 are factors of 20.

Fraction A number that shows part of a whole. Fractions are written with a numerator (top) number and a denominator (bottom) number separated by a dividing line. A unit fraction is a fraction that has a numerator of 1.

Improper fraction *See Fraction. See Mixed number.* A fraction that is greater than 1. An improper fraction has a numerator that is greater than the denominator, e.g. $\frac{8}{5}$, $\frac{11}{10}$ and $\frac{12}{2}$ are all improper fractions.

Irregular polygon A shape is irregular if the sides or angles are not all equal, e.g. a rhombus is irregular since, although the sides are all equal, the angles are not.

Inverse This is the opposite calculation. Addition and subtraction are opposites, e.g. the inverse of +10 is –10. Multiplication and division are also inverse operations.

Isosceles triangle A triangle with two equal sides. Isosceles triangles also have two equal angles opposite the equal sides.

Line symmetry Also known as reflective symmetry. A line that divides a shape into two equal parts. When the shape is folded on the line of symmetry, both halves fit exactly on top of one another, e.g. the dotted lines on this shape are all lines of symmetry.

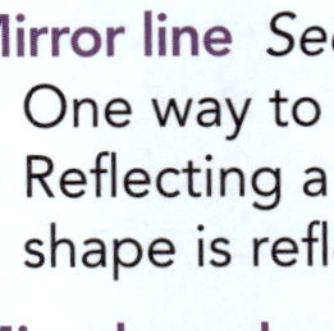

Mean The mean is a type of 'average'. To find the mean, add up the values and divide by the number of values added.

Mirror line *See Line symmetry. See Reflection.* One way to move shapes is to reflect them. Reflecting a shape means 'flipping' it over. The shape is reflected in the mirror line.

Mixed number *See Improper fraction. See Fraction.* A number that uses a whole number and a fraction, e.g. $6\frac{1}{3}$ or $10\frac{3}{8}$. A mixed number can be turned into an improper fraction, e.g. $2\frac{1}{5} = \frac{11}{5}$.

Multiple *See Factor.* The result of multiplying two whole numbers, e.g. 5 × 4 = 20, so 20 is a multiple of 5 and 4.

Net A 2-D representation of all the faces of a 3-D shape. Nets can be folded to make a 3-D shape, e.g. this is the net of a cube.

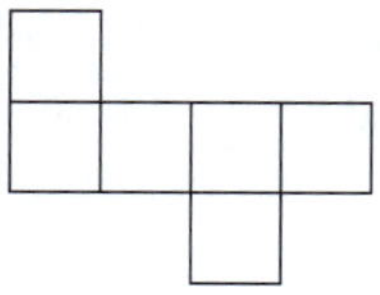

Numerator *See Denominator.* The number on the top of a fraction. The numerator shows how many parts are used in the fraction, e.g. in $\frac{3}{5}$ there are five fifths or parts needed to make 1 and this fraction uses three of the five parts.

Parallel Two or more lines that always stay the same distance apart. Parallel lines can be straight or curved. Small arrows drawn on two or more lines are a symbol that they are parallel.

Percentage A way of writing a fraction with a denominator of 100. The per cent symbol is % and it means 'out of a hundred', e.g. 99% means 99 per cent and is the equivalent of $\frac{99}{100}$.

Perimeter The length of the outside edge of a 2-D shape. It is measured in units of length such as metres or centimetres.

Perpendicular At right angles. Lines, sides of 2-D shapes or faces of 3-D shapes can be perpendicular.

Pie chart A circular graph that shows the proportion of different amounts by using the sectors of a circle.

Place value The value of a digit according to its position within a number, e.g. the 8 is worth 80 or eight tens in the number 2781 but it is worth 800 or eight hundreds in the number 3891.

Polygon Any straight-sided 2-D shape.

Prime number A number that only has two factors: itself and 1, e.g. 23 is a prime number as its only factors are 23 and 1. Note that 1 is not a prime number as it only has one factor, which is 1. Remember that 2 is the only prime number that is an even number.

Proportion *See Ratio.* Keeping the relative sizes of two or more numbers or amounts equal, e.g. 2 for every 5 is in proportion with 4 for every 10 and 6 for every 15. Proportion keeps ratios equal, e.g. 2:5 = 4:10 = 6:15.

Ratio *See Proportion.* Ratio compares the sizes of two or more numbers or amounts. Ratios are usually written with a colon and the word 'to' is said in place of the colon, e.g. 3:1:5 (say 'three to one to five').

Reflection *See Mirror line. See Line symmetry.* A shape that has been 'flipped' over in a mirror line. Remember, the shape should stay the same size but be reversed. Both the shape and its reflection must be the same distance from the mirror line.

Regular polygon A shape is regular if all the sides and all the angles are equal. Shapes where sides or angles are different are called irregular.

Right angle An angle that has 90°. A right angle is a quarter of a full turn.

Right-angled triangle A triangle with a right angle.

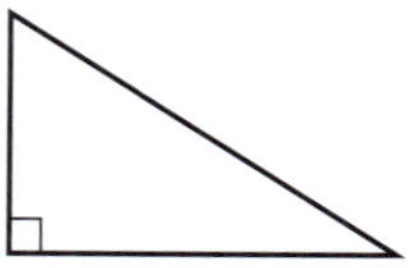

Rounding To give an approximate value to a number. Usually a number is rounded to a multiple of ten or a hundred, e.g. 4,638 rounded to the nearest hundred is 4,600.

Sequence A set of numbers that has been organised according to a rule, e.g. 5 9 13 17 21. This set of numbers is organised according to an 'add 4' rule.

Square number The result of multiplying a number by itself, e.g. 5 × 5 = 25 so 25 is a square number. The abbreviated way to write 5 × 5 is to write 5^2 (5 squared).

Trapezium A quadrilateral with one pair of parallel sides.

Translation A movement of a shape. Shapes are translated by 'sliding' them to the right or left, or up or down, often on a grid. Remember that a shape that has been translated does not change size or orientation (which way round it is).

Vertex A corner. The vertex of a 2-D shape is where two sides meet. The vertex of a 3-D shape is where the edges meet. If there is more than one vertex, use the word *vertices*.

***x* axis** The horizontal axis on a graph or grid, which usually runs through 0.

***x* coordinate** The value of a number or position on the *x* axis.

***y* axis** The vertical axis on a graph or grid, which usually runs through 0.

***y* coordinate** The value of a number or position on the *y* axis.

Answers

Number and place value

Place value of whole numbers (page 8)

1 400 (four hundred)
40,000 (forty thousand)
40 (forty)
400 (four hundred)
4,000,000 (four million)

Comparing and ordering whole numbers (page 9)

1 639,422 624,429 364,922 349,244 62,994

Rounding (page 10)

1 Any number from 8,450 up to 8,549 rounds to 8,500

Place value of decimal numbers (page 11)

1 $\frac{4}{100}$ (4 hundredths)
$\frac{4}{10}$ (4 tenths)
$\frac{4}{100}$ (4 hundredths)
4 (4 ones)
$\frac{4}{10}$ (4 tenths)

Negative numbers (page 12)

1 24
2 15
3 14, – 2

Number – Addition, subtraction, multiplication and division

Addition (page 13)

1 77,761
2 2,925
3

	6	3	[4]	5
+		[7]	2	[3]
	7	0	6	8

Subtraction (page 14)

1 7,642
2 £128,171
3

	8	[8]	5	9
+	[6]	3	4	[6]
	2	5	1	3

Multiplying and dividing by 10 and 100 (page 15)

1 a) 15,320 b) 15.32
2 a) 10 b) 6,300
3 a) 9.5 b) 77

Multiples and factors (page 16)

1 1 × 72; 2 × 36; 3 × 24; 4 × 18; 6 × 12; 8 × 9 (in any order)
2 Numbers placed as shown

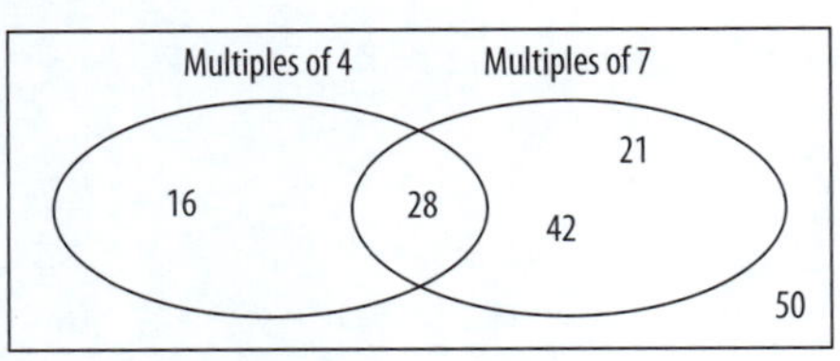

Multiplying by larger numbers (page 18)

1 5,112
2 12,828
3 10,464
4 195,372
5 5,350

Square numbers (page 19)

1 a) 16 b) 81
2 9; 16; and 25
3 12
4 6

Short division (page 20)

1 103
2 652
3 986 r5 (accept 986 $\frac{5}{8}$ or 986.625)
4 218

Long division (page 21)

1 27
2 95
3 152

Prime numbers (page 22)

1 11; 13; 17; 19
2 No, Kate is incorrect. Explanation to show that 2 is a prime number because it only has one pair of factors – itself and 1.

Number – Fractions, decimals and percentages

Fractions of amounts (page 23)

1 a) 18 b) 16 c) 13 d) 36
2 a) 60 b) 40

Mixed numbers (page 24)

1 a) $\frac{5}{2}$ b) $\frac{51}{10}$ c) $\frac{17}{5}$
2 Circle $\frac{16}{6}$ and $\frac{24}{9}$

Equivalent fractions (page 25)

1 a) $\frac{1}{5} = \frac{2}{10}$ b) $\frac{5}{9} = \frac{15}{27}$ c) $\frac{2}{3} = \frac{6}{9} = \frac{14}{21}$
2 a) $\frac{5}{6}$ b) $\frac{4}{5}$ c) $\frac{5}{7}$ d) $\frac{1}{2}$
3 $\frac{15}{20}$ and $\frac{14}{20}$ (accept $\frac{30}{40}$ and $\frac{28}{40}$ or other equivalents)

Adding and subtracting fractions (page 26)

1 a) $\frac{4}{7}$ b) $\frac{2}{6}$ (accept $\frac{1}{3}$) c) $\frac{7}{8}$
2 a) $1\frac{3}{8}$
b) $4\frac{7}{10}$

Fractions and their decimal equivalents (page 27)

1 a) 0.06 b) $\frac{55}{100}$ (accept $\frac{11}{20}$) c) 0.6
2 a) 0.25 b) 0.4 c) 0.75 d) 0.2

Adding and subtracting decimals (page 28)
1 £61.03

Multiplying decimals (page 29)
1 3.5
2 4.8
3 0.36
4 1.2
5 0.4
6 6
7 0.09
8 9

Percentages as fractions and decimals (page 30)
1 $\frac{75}{100}$ (accept $\frac{3}{4}$), 0.75
2 $\frac{6}{10}$ (accept $\frac{3}{5}$), 60%
3 0.3, 30%

Finding percentages (page 31)
1 16
2 30
3 90
4 100
5 137.5

Ratio and proportion

Ratio and proportion (page 32)
1 30
2 250 ml

Algebra

Algebra (page 33)
1 230 minutes
2 2.5 (accept $2\frac{1}{2}$)

Sequences (page 34)
1 a) 50 59 Rule is add (+) 9
b) 1 –3 Rule is subtract (–) 4
c) 90 130 Rule is add (+) 40
2 1,500; 4,500; 7,500

Solving equations (page 35)
1 a) Accept any correct numbers, e.g. 40 + 1 = 90 – 49
b) Accept any correct answer,
e.g. 10 × **1** = 200 – **190**; 10 × **2** = 200 – **180**; 10 × **3** = 200 – **170**

2 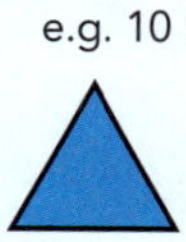 = 25

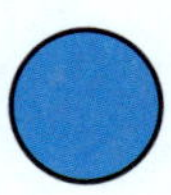 = 15

Measurement

Length (page 36)
1 2 m indicated only
2 0.97

Mass (page 37)
1 2.5 kg indicated only
2 750

Capacity (page 38)
1 50,000
2 12
3 375

Money (page 39)
1 £2.25 (accept 225p)
2 £12.35
3 400
4 £72.25

Time (page 40)
1 2:10 p.m.
2 19:45
3 50

Time problems (page 41)
1 80
2 6:55 p.m. (accept 18:55)
3 50

Perimeter (page 42)
1 46
2 11

Estimating the area of irregular shapes (page 43)
1 11 (+/– 2)

Area by formula (page 44)
1 121
2 45

Geometry – Properties of shapes

Drawing lines and angles (page 45)
1 Shape drawn accurately and to scale

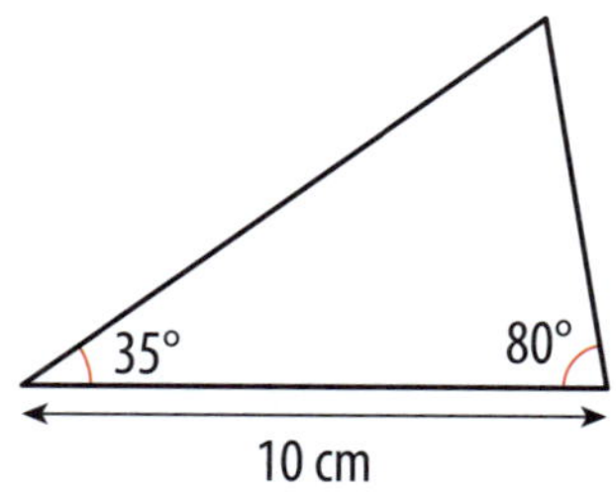

Accept slight inaccuracies:
accept lines +/– 2 mm and 80° angle +/– 2°

2-D shapes (page 47)
1

	No right angles	1 right angle	2 right angles	3 right angles	4 right angles
No pairs of equal sides	3				
1 pair of equal sides	2	1	6		
2 pairs of equal sides			5		4

3-D shapes (page 49)
1 d)

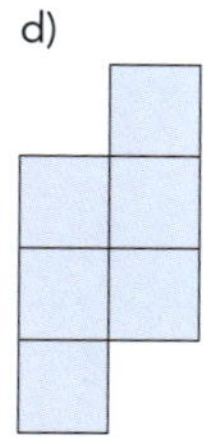

2 a) cylinder b) cuboid c) triangular prism

3 Nets indicated as shown

Angles and degrees (page 50)

1 $a =$ 195

Angles in triangles (page 51)

1 90 and 40

Geometry – Position and direction

Coordinates (pages 52–53)

1 (3,4)
2 (1,5)
3 (8,3)

Translations (page 55)

1–2

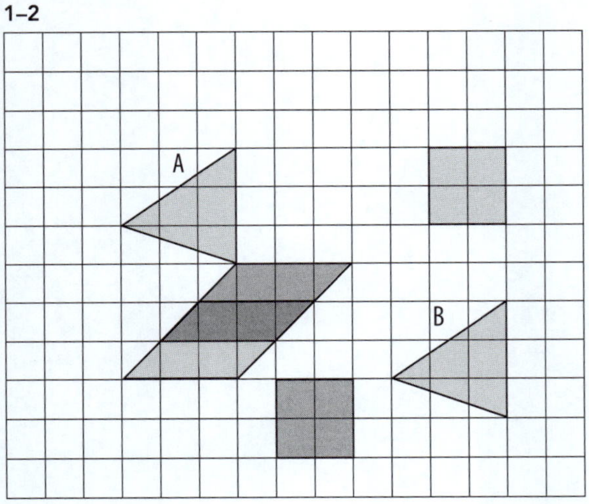

3 7 right, 4 down

Reflections (page 56)

1

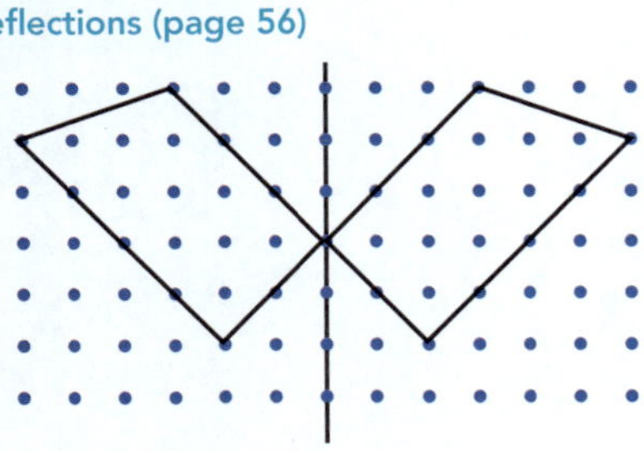

2

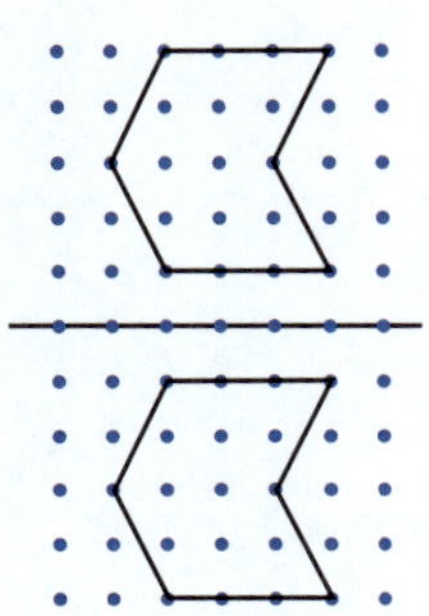

Reflective symmetry (page 57)

1 square: 4; parallelogram: 0; regular hexagon: 6

Statistics

Tables (page 58)

1 369 children
2

	London	Bristol	Belfast	Manchester	**Total**
Week 1	32	15	**16**	8	**71**
Week 2	**36**	**12**	16	**12**	76
Total	68	27	**32**	**20**	147

Pictograms (page 59)

1 tuna
2 70
3 180

Bar charts (page 60)

1 Thursday and Friday
2 £10
3 £210

Pie charts (page 61)

1 $\frac{1}{8}$
2 $\frac{3}{4}$
3 25

Line graphs (page 62)

1 3:00 p.m. (accept 15:00)
2 10

Mean (page 63)

1 9
2 £22.50
3 27

Problem solving and reasoning

Whole number problems (page 64)

1 16,464
2 25,000

Problems with fractions, decimals and percentages (page 65)

1 3

Measurement problems (page 66)

1 Acceptable answers: 74, 40, 30, 26 (accept other answers that involve lengths and widths that involve fractions of decimals, if supported by accurate calculation)
2 Acceptable answers: 19, 36, 51, 64, 75, 84, 91, 96, 99 (accept other answers that involve lengths and widths that involve fractions of decimals, if supported by accurate calculation)

Shape problems (page 67)

1 $p = 60$; $q = 120$

Reasoning (page 68)

1 An explanation which recognises that 10% of 95 is not a whole number, e.g. 10% of 95 is $9\frac{1}{2}$ and you cannot have $9\frac{1}{2}$ people.